MICROPHONES&MIXERS

2ND EDITION

Bill Gibson

HAL LEONARD RECORDING METHOD, BOOK 1

Hal Leonard Books

An Imprint of Hal Leonard Corporation

Second edition published in 2011 by
Hal Leonard Books
An Imprint of Hal Leonard Corporation
7777 West Bluemound Road
Milwaukee, WI 53213

Trade Book Division Editorial Offices
33 Plymouth St., Montclair, NJ 07042

First edition published in 2007 by Hal Leonard Books

Printed in the United States of America

Book design, illustrations, and DVD audio and video production by Bill Gibson
Front cover photo courtesy of Steve Burdick/Westlake Recording Studios & Technical Services
Library of Congress Cataloging-in-Publication Data is available upon request.
ISBN 978-1-45840-296-7
www.halleonardbooks.com

Acknowledgments

To all the folks who have helped support the development and integrity of these books. Thank you for your continued support and interest in providing great tools for us all to use.

- Acoustic Sciences, Inc.
- AKG
- Antares
- Apple Computer
- Audio-Technica
- Avid
- Jeff Busch: Percussion
- Geoffrey Castle: Violin
- Jamie Dieveney: Vocals, songwriting, friendship
- Eugene Bien: Piano
- Doug Gould: Worship MD
- Faith Ecklund: Vocals, songwriting, inspiration
- Gibson Guitars
- Glimpse: Josh and Jason, you rock!
- Steve Hill: Drums
- IK Multimedia
- JBL Pro: Peter Chaikin, Joseph Wagoner, and Alfred Reinprecht
- Mike Kay: Ted Brown Music
- London Bridge Studios: Geoff Ott, Jonathan Plum, Stephen Michael Hogan
- Thomas Marriott: Trumpet
- Monster Cable
- John Morton: Guitar and stories
- Native Instruments
- Ben O'Brien: Production Assistant
- Primacoustic Studio Acoustics
- PreSonus
- Radial Engineering
- Robbie Ott
- Roger Wood
- Royer Labs
- Sabian Cymbals
- Shure
- Spectrasonics
- T.C. Electronic
- TC Helicon
- Taye Drums
- Universal Audio
- Waves
- Yamaha
- Paul Zimmerman: Assistant

Contents

vi

Microphones & Mixers..by BILL GIBSON

Chapter 6—Processors 227

Chapter 7—Effects 267

Audio and Video Examples

xii

Microphones & Mixers...by BILL GIBSON

xiv

Microphones & Mixers...by BILL GIBSON

xvi

Preface

Welcome to the Hal Leonard Recording Method by Bill Gibson. The material in this six-book method has been in development for many years. In the early 1980s I was engineering and producing in the Seattle area and at the same time teaching audio engineering classes. My classes ran fairly continuously, and I usually had two or three classes happening at once. It was a fun time in the industry because it was just prior to the home studio revolution, but the music industry was booming and there was a lot of excitement involved in just being in a recording studio.

After several years of teaching eager students of all ages, I collaborated with Bob Sluys, a friend from college, to begin to put some materials together that could help students around the world. We produced one of the first videos about audio recording. It really turned out well considering that it was shot in my family room and that we used Bob's camera and my gear. He had just moved back from LA and had developed his videography skills to the level where we could put together a nice-looking product for not much money. Editing needed to be done in a commercial facility—the desktop video revolution hadn't begun yet—but other than that, it was guerilla video all the way. It was a great experience just putting the material I had developed over the years together. I discovered that my constant exposure to students, my quest to answer their questions, and my desire to provide information that could help them succeed had resulted in a substantial arsenal of very pragmatic explanations for a lot of recording principles.

To make a long story short, I eventually ended up releasing the AudioPro Home Recording Course through MixBooks in 1996. Over the years I've been fortunate to to publish a lot of material, and the refinement process hasn't stopped. In each new book I've covered new topics or fine-tuned existing ones—this recording method is the culmination of a long journey. I sincerely hope it helps you produce the kind of recordings that come straight from your soul and that it helps you gain a high level of proficiency and professionalism.

There are six books in this method. They follow a logical progression through modern audio recording and production techniques and considerations. If you follow the material carefully, listen to the audio examples, watch the video examples, and practice on your own equipment, you will become a respectable engineer. You'll be able to finely craft individual sounds, you'll know how to build a mix, and you'll have a solid insight into a broad range of recording techniques and production concerns.

The audio arts—whether music, speech, sound design for visual arts, or archival—are on one hand a beautiful art form and on the other hand a very demanding and meticulous technical undertaking. It's important that we are driven by creative motivation; it's also important that we expand our technical understanding to support and extend our creative vision.

By design, this book leads you through the essential ingredients in audio recording: core equipment, the building blocks of all the technical and creative aspects of audio recording, and demonstrations that give you visual and aural instruction. Follow along on the accompanying DVD-ROM or online media. Each audio and video example is clearly noted and is meant to be experienced in conjunction with the reading.

· Audio Examples

Audio Examples are indicated like this.

· Video Examples

Video Examples are indicated like this.

The audio industry is ever changing. The mixer I use in my project studio is capable of performing audio gymnastics that were not possible at any price not that many years ago. The fact is, relative to features, most computers today substantially outshine the most extravagant professional recording consoles of the '80s and early '90s; that's all within the computer itself, without any mixer at all. Our creative options are enormous. We can bend, shape, mold, stretch, tune, distort, restore… the list goes on and on. With few exceptions, whatever you can imagine musically, you can achieve.

Technology has given us the ability to create in a new and exciting way. With that in mind, we all need to remain focused on what has drawn us to music and audio. Most of us just love music and appreciate a well-recorded, artistic, and emotional performance. That's the bottom line—capturing beautiful art. Beauty is in the eye—or in our case, the ear—of the beholder. I love the fact that each genre has a distinct personality. I also appreciate the fact that any given genre has the potential to thrill one group of people while repulsing another. Yet, there is a common ground—appreciation of art. This universal appreciation, in its purest form, supersedes personal taste. In other words, good music is good music. Let your passion for great music and audio guide your attention to technical details.

This book teaches you specific concepts and techniques that will improve the overall quality of your audio recordings so that they will viably compete with music you hear on your favorite recordings. Herein are set out basic concepts and terms as they apply to our fundamental tools. Once you've grasped the principles and concepts I describe, you should immediately put them into practice. Each new concept or technique is mentioned to help you be more productive while supporting and enhancing your creative flow.

This material will help you get sounds that are competitive. Persevere! Keep fine-tuning your craft. If you're serious about audio as a career or if you're doing music just for the fun of it, this book is for you. The techniques described will help you make better use of your recording time. Your music will greatly benefit from your deeper understanding of the studio as a musical tool.

The primary tools of the musical trade, for both the professional and the amateur, are available everywhere, right off the shelf. Technology is more affordable now than ever before. With a mixer, some keyboards or guitars, a microphone, and a computer, almost anyone can create a solid musical work that either can be completed at home or can be polished off in a professional recording studio. With some motivation, imagination, and education, you can (in your own hometown) make your music a financially, emotionally, and artistically profitable venture.

We're going to approach recording from a musical perspective. You'll study recording examples that fit real musical situations, and you'll learn solutions to common problems that will help you enhance your music. In doing so, you'll establish a base of knowledge. As your guide through this material, I'll keep the focus on details that make a difference in understanding how technology serves music, rather than how music serves technology. There are certainly plenty of places for a guy or gal who walks into a recording session with a voltmeter in one hand and an oscilloscope in the other—in fact, they're essential. However, I tend to show up with a guitar in one hand and a mouse in the other.

THE ACCOMPANYING DVD-ROM AND ONLINE MEDIA

All media examples can be found on the accompanying DVD-ROM and online at www.hlrmonline.com. Stream the examples online for super-easy access or, if you prefer, import all of the full-resolution media from the DVD-ROM directly into iTunes or any other full-featured media player for easy playback. This is especially convenient when you're on the go with your laptop, tablet device, or smartphone and unable to connect to the Internet.

All Audio and Video Examples are produced specifically to support the concepts and principles presented in the printed text, with the examples clearly marked in each chapter. The Audio Examples demonstrate many of the concepts that are explained in the text and accompanying illustrations. The Video Examples show you specifically how to build your skills as a sound operator. They are very powerful and they're produced with your education in mind. You won't find a lot of rapid-motion, highly stylized shots; you will find easy-to understand instructional video that is edited for optimal instruction and learning.

QR CODES

Throughout this book, you'll see an occasional QR code. These codes are actually digital links to supporting websites; they're designed to be read by a QR code reader on a smartphone, such as the iPhone or Android or on a tablet, such as the iPad or Xoom. Simply download a free QR Code Reader app, show the code to the smart device's camera, and the supporting link will automatically show up in its browser.

QR codes are a convenient way to quickly delve deeper into the supporting material used to help create this work. Once you have your QR code reader, let your smart device scan this QR code to be taken directly to the support site for the Hal Leonard Recording Method.

CHAPTER TESTS

Each chapter ends with a 20-question test. These tests are designed to help reinforce the information you've learned. Retake the tests until you're sure that you understand the concepts presented in each chapter. The knowledge presented throughout this method is cumulative. If you understand each chapter, you will be able to function well in the world of high-quality audio.

CONCLUSION

A number of years ago the home studio was shunned by large commercial studios. In this day and age, it's a fact that very many commercially successful major music projects are, at the very least, partially recorded in someone's home studio. Creativity has become the main issue. Whether you're in a large commercial facility or in a bedroom studio, keep your standards high and crank out the hits!

Sound Theory

It's very important that we understand the principles of sound in order to accurately capture music, speech, sound effects, or even noise. The sound source helps define the tools we should use to record it. Most of the time we strive to preconceive the sonic impact of the recording. Other times, we stumble across a sound that inspires a complete redirection of the artistic process. In either case, a thorough understanding of your recording tools is essential.

Any time you're recording an acoustic instrument, listen to the instrument first. Stand beside the musician and hear what he or she hears. Listen to the sound of the instrument, or voice, decaying in the room. Stand close and move away. Assess the sonic differences in the acoustic space. If you really want to capture the true essence of the sound, you'll need to make excellent decisions about where the instrument is placed in the room, what microphone you'll use, and where you'll place it in relation to the instrument. If you want to capture something other than the true sound of the instrument or voice, you'll need to be fully aware of the options available to you, and you'll need to be able to use them in a creative and artistically supportive manner.

The information we're about to cover is fundamental to the understanding of sound. Study it carefully. It will help you make great recordings of great music.

2

Microphones & Mixers... by BILL GIBSON

CHARACTERISTICS OF SOUND

Sound is energy that travels through air. Air molecules move in relation to the sound that moves them. When something vibrates, like a drum, string, vocal chord, etc., it affects the air around it. The air responds to the vibrations directly, contracting and expanding as the vibrating material completes its cycle of vibration. These vibrations cause continuous variations in the existing air pressure.

· ·Video Example 1-1

Comparison of Wavelengths

Waves

Visualize sound in air like a wave in water. Any sound creates a disruption in the stillness of air, just like dropping a rock in a lake creates a disruption in the stillness of water. In fact, sound is referred to as sound waves because of this simple concept. As with so many concepts in art or science, the basic principles are easy to understand. Most complex theories and concepts can be stripped down to a fairly reasonable string of simple ingredients.

As you watch the waves in water travel, they minimize in size until they disappear, unless they reach an outer boundary, in which case they reflect back

Sound Wave Reflections

Sound waves move in air like waves move in water. Interactions occur between waves and their reflections spherically in all directions from the sound source.

Acoustic Reflections Combine with the Source

Sound emanates omnidirectionally from the source. However, when we place a microphone in front a source we also get the reflections off each surrounding surface, combined together at the mic.

Wall

Source

Wall

toward the center. The amount of reflection depends on the energy at the source and the distance to the boundary. As the waves radiating from the source meet the waves rebounding from the boundary, it's easy to see the waves interacting and influencing each other's shape and size. That's exactly what happens to audio in an enclosed space. The reflections combine with the source audio, each influencing the other. This is the reason any given instrument or voice takes on a different character, or timbre, depending on the space it is in.

Our perception of sound is directly related to the waves in air. Consider that a lack of sound is completely still air—mighty difficult to find, but let's just imagine it for the sake of understanding. As soon as there is vibration by anything in the still air (normal atmospheric pressure), waves begin. In relation to still air, each wave contains a crest and a trough.

When the wave touches our eardrum, the membrane vibrates in sympathy with the source, being pushed in by the crest and pulled out by the trough. This is explained by the principle of sympathetic vibration. When a sound wave strikes a body, which will naturally produce the same wave, the vibration of the body is called sympathetic vibration.

4

Microphones & Mixers .... by BILL GIBSON

Crest and Trough

Audio waveforms consist of a series of crests and troughs that push and pull on the eardrum. This simplest waveform, called a sine wave, has the smoothest curve and a sound similar to a flute.

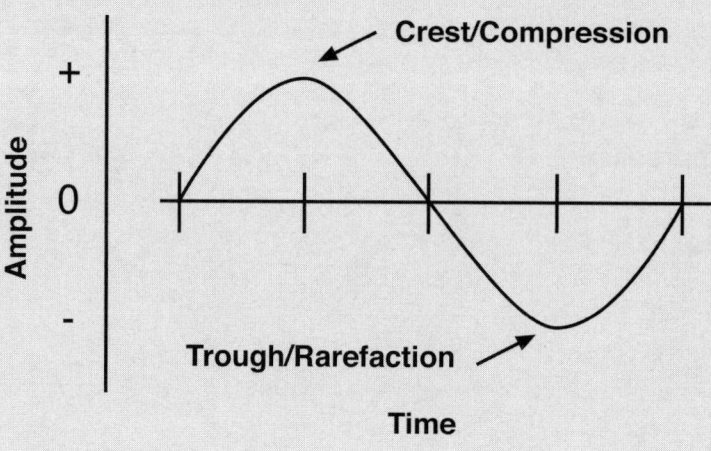

The simplest of sound waves is called a sine wave. When we chart the rise and fall of the crest and trough, the sine wave is perfectly smooth and provides an excellent illustration of the basic aspects of sound. To illustrate a sound wave we consider a straight line as still air. As the wave ascends above the line, creating a crest, our eardrum is pushed in—this is also called the compression portion of the cycle. As the wave descends below the line, creating a trough, the eardrum is pulled out—this is called rarefaction. Compression causes the air molecules to bunch together, causing an increase in air pressure. Rarefaction is the result of the air molecules filling in behind the compression, resulting in a decrease in air pressure.

Keep in mind that individual air molecules don't travel from the sound source to the listener's eardrum. Sound is merely causing a chain reaction, which moves air molecules back and forth, causing the air molecules they're touching to move back and forth, and so on, until the air molecules that touch your eardrum initiate its movement. In Seattle, at the football games, we all get a big kick out of doing "The Wave," where a chain reaction flows all around the stadium. As soon as the person next to you stands up, you stand up, and as they're sitting down, you sit down. A huge wave moves all around the stadium, which looks really cool and for some strange reason makes you feel good about life. No one has to run around

Push and Pull on the Eardrum

The pinna focuses sound toward the eardrum—it is fundamental in localization of sound. Compression and rarefaction are channeled into the ear canal, where the changing air pressure vibrates the tympanic membrane, which begins the process of sending corresponding electrical impulses to the brain. Ah yes … the tree did fall.

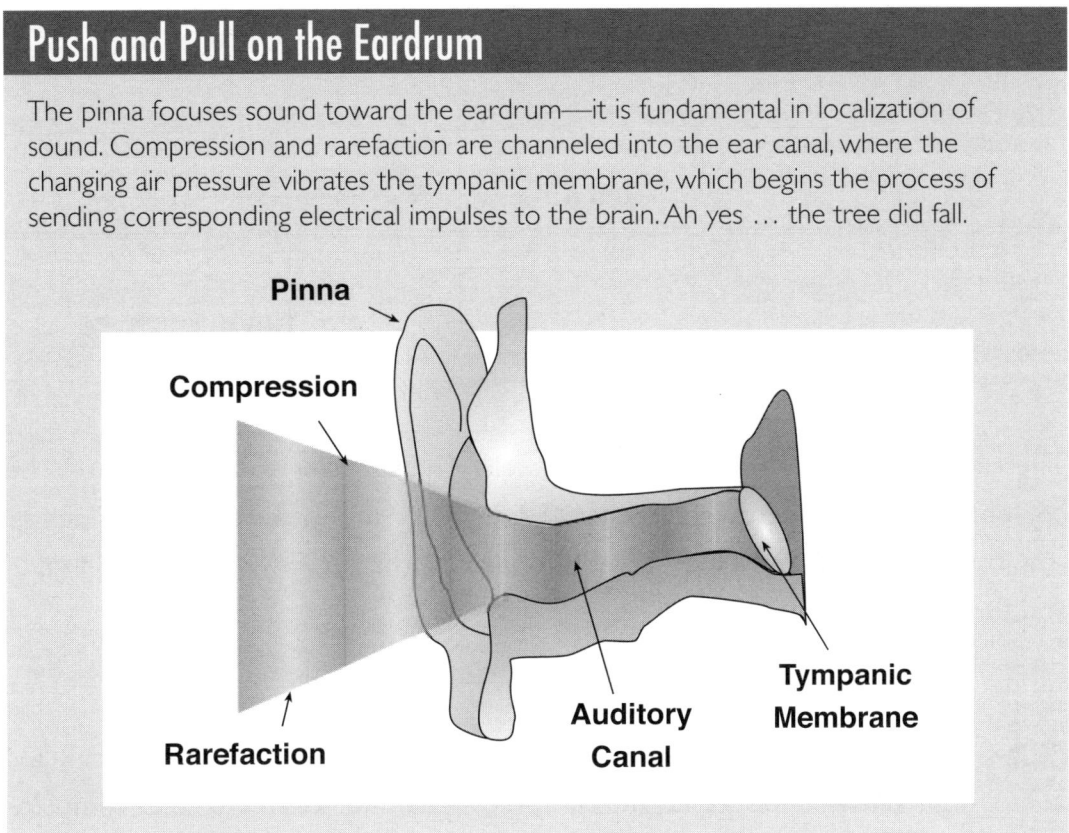

the stadium, but the wave makes it all the way around. That's how sound transfers through air.

Speed

In normal atmospheric conditions sound transfers through air relatively slowly, at the rate of about 1,126 feet per second (about 340 meters per second, 30 cm per millisecond, or just over one foot per millisecond.) Elevation, temperature and humidity affect the speed slightly. The speed of sound in air is determined by the conditions of the air, not by waveform characteristics like amplitude, frequency, or wavelength. There are plenty of formulas available on the Internet that take all factors into consideration for the speed of sound. For the purpose of our illustrations and calculations, 1,126 ft./sec. will usually suffice.

Cycles

When a source has completed one crest and one trough it has completed one cycle. This represents a push and pull on the eardrum—one compression and one rarefaction of air molecules. We quantify position during the cycle in degrees.

360-Degree Wave

From beginning to end, a complete cycle of any waveform is quantified as 360 degrees. Halfway through the wave cycle is 180 degrees, one quarter of the way through is 90 degrees, etc.

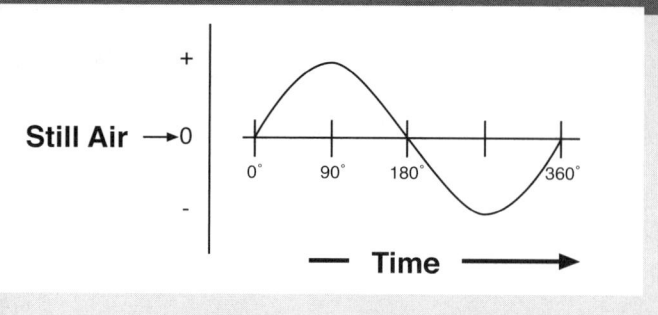

There are 360 degrees in one complete cycle. The zero degree mark denotes the beginning of compression (the crest); 360 degrees denotes the end of rarefaction (the trough). 180 degrees marks the midpoint of the cycle, where the crest ends and the trough begins.

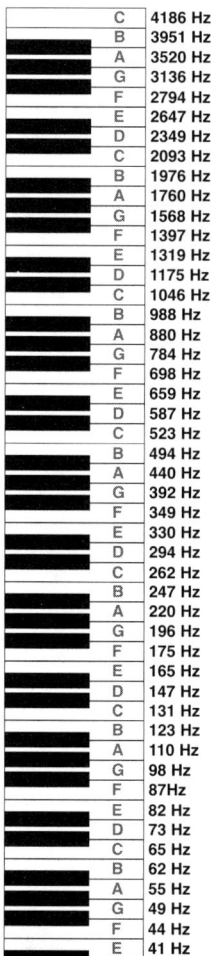

C	4186 Hz
B	3951 Hz
A	3520 Hz
G	3136 Hz
F	2794 Hz
E	2647 Hz
D	2349 Hz
C	2093 Hz
B	1976 Hz
A	1760 Hz
G	1568 Hz
F	1397 Hz
E	1319 Hz
D	1175 Hz
C	1046 Hz
B	988 Hz
A	880 Hz
G	784 Hz
F	698 Hz
E	659 Hz
D	587 Hz
C	523 Hz
B	494 Hz
A	440 Hz
G	392 Hz
F	349 Hz
E	330 Hz
D	294 Hz
C	262 Hz
B	247 Hz
A	220 Hz
G	196 Hz
F	175 Hz
E	165 Hz
D	147 Hz
C	131 Hz
B	123 Hz
A	110 Hz
G	98 Hz
F	87Hz
E	82 Hz
D	73 Hz
C	65 Hz
B	62 Hz
A	55 Hz
G	49 Hz
F	44 Hz
E	41 Hz
D	37 Hz
C	33 Hz
B	31 Hz
A	27 Hz

Frequency

The frequency of sound quantifies the number of times a wave completes its cycle in one second. In relation to pitch, higher frequencies complete more cycles each second. Frequency is expressed in Hertz (Hz) or cycles per second (cps). 100 Hertz, or 100 cps, represents a waveform that completes its cycle 100 times each second. 1,000 Hertz equals 1 kilohertz (kHz). As frequency increases, it's common to refer to v multiples of kilohertz. 2,500 Hertz is typically referred to as 2.5 kHz; 5,100 Hertz = 5.1 kHz. In common usage, we often refer to kilohertz simply as "k." For example, it's common to say, "I boosted the vocal track at 4 k."

Frequencies and Pitch

Each musical note is related to a specific sound wave frequency. In the illustration to the left, we see whole number frequencies that, though they've been rounded off, indicate the way the frequency range of most music relates to pitch.

The specified number of Hertz indicates the fundamental frequency—the sine wave that defines the specific pitch and octave. The fundamental is often referred to as f0.

Instrument Ranges Compared to the Piano Keyboard

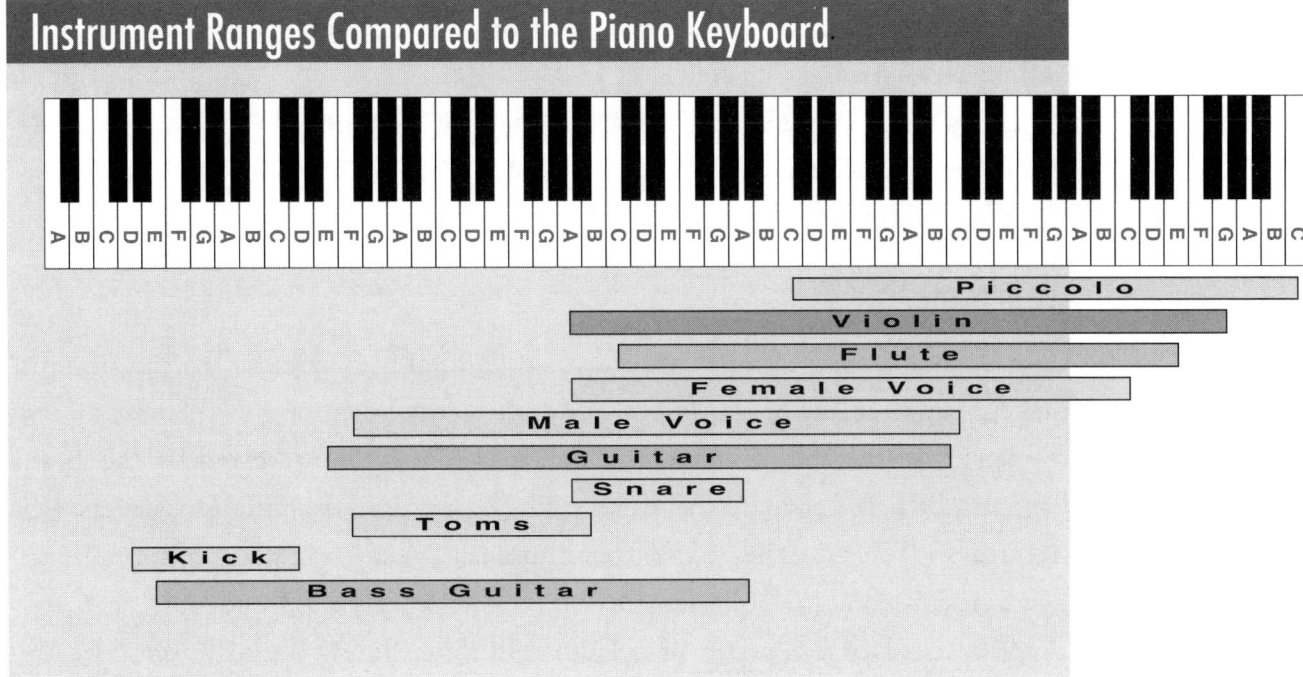

The frequency response range of the human ear is roughly from 20 Hz to 20 kHz. Brand-new ears, like those in a baby, tend to be able to hear frequencies above 20 kHz, sometimes approaching 23 kHz. Old tired ears, like those found in many musicians, probably don't hear high frequencies as well as they used to.

Pitch

An octave on the piano is the distance from a note to the next note of the same name. From middle C to the C above middle C is one octave. Mathematically, an octave above any pitch is twice the frequency of the pitch. On octave below any pitch is half the frequency of the original pitch.

Each pitch (note on a piano, guitar, trumpet, etc.) has a specified fundamental frequency within our system of tonality. We use a 12-tone system (12 notes per octave), tuned in a specific, tempered way. There are many other tonalities throughout the world, utilizing different numbers of notes per octave.

Middle C on the piano has a funda-mental frequency of 262 Hz. The A above middle C is often used as a standard tuning reference, and is called A 440—indicating a fundamental frequency of 440 Hz.

The frequency that defines the pitch name is called the fundamental frequency. In reality there is much more to a note than its fundamental frequency. Aspects of the sound wave called harmonics, overtones, and partials determine the individual character of a sound. The fundamental only determines the name and octave of a note.

As a point of reference, the lowest note on a standard 88-note piano keyboard has a fundamental frequency of 27 Hz. The highest note on the piano has a fundamental frequency of 4,186 Hz.

Wavelength

Low frequencies have longer waveforms than high frequencies. The physical distance in air from the beginning of one cycle to the beginning of the next cycle is the length of the sound wave. The wavelength is often indicated by the Greek letter lambda. To calculate the wavelength (λ), we use a formula consisting of the frequency (f) specified in cycles/second, and the speed, or velocity, of sound specified in feet/second (v). Wavelength (λ)= Velocity (v) ÷ the frequency (f). $\lambda = v/f$

To calculate the length of a 1,000 Hz tone, simply plug the variables and constants into the formula. The constant is the speed of sound (v), clocking in at 1,126 feet/second. In this case the wavelength (λ) = 1,126 (v) ÷ 1,000 Hz Therefore, $\lambda = 1,126/1,000$, which equals 1.12 feet long.

Wavelength - Proportional Length Comparison

This illustration would be a much more impressive wavelength comparison if it were full scale. The lowest discernible pitch (20 Hz) is 56 feet long, just about the length of four Ford Explorers; the highest discernible pitch (20 kHz) is barely longer than half an inch.

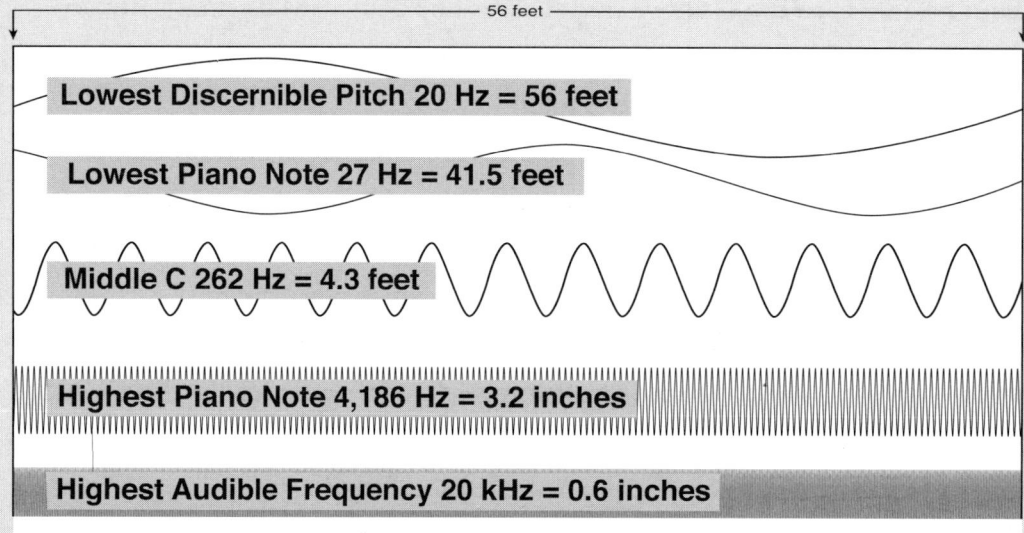

The lowest note on the piano (27 Hz) is calculated $\lambda=1,126/27$. The result of this equation indicates a wavelength of about 41.5 feet. The highest note on the piano (4,186 Hz) is calculated $\lambda=1,126/4,186$. The result of this equation indicates a wavelength of just over three inches (.27 feet).

Our understanding of wavelength is crucial to our understanding of acoustics and how sound reacts to, and interacts with, its environment. There are some situations where we need to calculate the frequency of a specific wavelength. This is a simple task of cross-multiplication in which we find that $f=v/\lambda$. These equations will be important in our studies on basic acoustics.

Amplitude

Amplitude expresses the amount of energy in a specified sound wave. Charted on a graph, a waveform with twice the amplitude has a crest that rises twice as high and a trough that dips twice as low.

Amplitude only compares the energy of a sound wave. It's a simple comparison: a waveform with maximum amplitude that's 2.5 times higher than another waveform contains 2.5 times the energy. The unit commonly used to quantify amplitude is dB SPL (decibels Sound Pressure Level). This is an objective scale based on mathematical logarithmic comparisons expressed as $decibel=10 \log10(P1/P2)$.

Any increase in amplitude indicates an increase in volume. However, the correlation is not always direct—twice the amplitude does not always indicate twice the volume. The relationship is dependent on the frequency and loudness.

Loudness

Loudness is a sound characteristic that involves the listener—it is a perceived characteristic that can be charted and averaged, but it's not simply a mathematical calculation. The common unit, used to quantify loudness, is the phon. Loudness is a subjective, perceptual aspect of sound.

The human ear is not equally sensitive to all frequencies. In fact, as amplitude varies so does the frequency response characteristic of the ear. The ear is most sensitive between 1 and 4 kHz. This frequency range just happens to contain the frequencies that give speech intelligibility, directional positioning, and understandability. Hmmm … it's almost like it was designed that way. In

The Loudness of Everyday Life

Examples of everyday noise levels in dB SPL

Weakest sound heard	0 dB
Normal conversation (3 - 5')	60 - 70 dB
Telephone dial tone	80 dB
City traffic (inside car)	85 dB
Train whistle at 500'	90 dB
Subway train at 200'	95 dB

Sustained exposure may result in hearing loss at these levels

Possible hearing loss	90 - 95 dB
Power mower	107 dB
Power saw	110 dB
Pain begins	125 dB
Pneumatic riveter at 4'	125 dB
Jet engine at 100'	140 dB
Death of hearing tissue	180 dB
Loudest sound possible	194 dB

fact, as the amplitude decreases, our ears become dramatically more sensitive in this frequency range.

So, yes, there is a difference between amplitude and volume. They are very similar at a certain point, though. Two scientists at Bell Laboratories in 1933 charted a survey of perceived volume. They compared actual amplitude to perceived volume throughout the audible frequency range (x-axis) and the accepted range of normal volume (y-axis).

The results of their survey involved generating pure tones through the audible frequency and volume spectrum at a specific amplitude, then asking numerous individuals to subjectively identify if the sound was louder or softer than the reference. Their survey, referred to as the Fletcher-Munson Curve, is a very visual representation of why music sounds fuller at loud volumes and thinner at soft volumes.

Each curve on the graph represents perceived constant volume throughout the audible frequency range. This, for example, shows us that in order to perceive 70 phons of loudness at 1,000 Hz requires 70 dB SPL (amplitude). However, in order to perceive 70 phons at 50 Hertz, 80 dB SPL is required. At 10 kHz, to perceive 70 phons, a similar 10 dB SPL boost is required.

As dB SPL decreases, the contrast becomes even more extreme between loudness and the actual amount of dB SPL required. At 20 phons, 20 dB SPL is equal to 20 phons. In contrast, at 50 Hz almost 65 dB SPL is required to maintain the perceived 20 phons.

Analysis of the Fletcher-Munson curve points us to the dB SPL range at which the human ear is most accurate throughout the audible frequency spectrum. Notice that between roughly 700 Hz and 1.5 kHz, phons are essentially equal to dB SPL at all volumes. Also, notice that at the center of the graph is where more often than not dB SPL is most similar to phons.

From this graph it is generally held that the most sensitive frequency range is from 1–4 kHz, although the graph might indicate an extension of that range from about 700 Hz to 6 kHz or so. Since this is a subjective study, some generalities apply but it is obvious where the consistencies and trends are.

For our recording purposes, it is constructive to find the flattest curves on the graph. A curve with less variation indicates a volume where the human ear's response most often matches loudness to dB SPL—the level where the most accurate assessments can be made regarding mix and tonal decisions.

Government Regulations on Exposure to Loud Sounds

OSHA Daily Permissible Noise Level Exposure

The Occupational Safety and Health Administration (OSHA) is part of the U.S. Department of Labor. This organization has studied and prescribed maximum sound pressure levels in the workplace, in relation to the number of hours per day the worker is exposed. These guidelines are useful to help audio engineers guard against permanent hearing loss.

Hours per day	Sound level
8	90 dB
6	92 dB
4	95 dB
3	97 dB
2	100 dB
1.5	102 dB
1	105 dB
.5	110 dB
.25 or less	115 dB

12

Microphones & Mixers .... by BILL GIBSON

The LOUDNESS button on your stereo is an example of compensation for the fact that it takes more high and low frequencies at a low volume to perceive equal loudness throughout the audible spectrum.

The most consistent monitor volume for our recording purpose is between 85 and 90 dB SPL, according to the Fletcher-Munson Curve. Notice on the graph that the 80 and 90 phons curves are the flattest, from 20 Hz to 20 kHz.

There are a few different devices available to help you quantify specifically how loud, in dB SPL, you have your system set. The simplest and least expensive way to assess dB SPL is with a handheld decibel meter. They are available at most home electronics stores and, depending on features and manufacturer, typically range in price from about $40–$300. Most of these instruments offer A- and C-weighting, along with slow (average) and fast (peak) attack times.

The Fletcher-Munson Curve of Equal Loudness

This graph plots results from a survey that relates amplitude (dB SPL) to perceived volume. This curve is valuable because it highlights the frequency response characteristic of the human ear. Since amplitude is a quantifiable energy level and loudness is a subjective characteristic, based on the listener's opinion, there's no better way to discover perceived volume than to ask human beings and then chart the results.

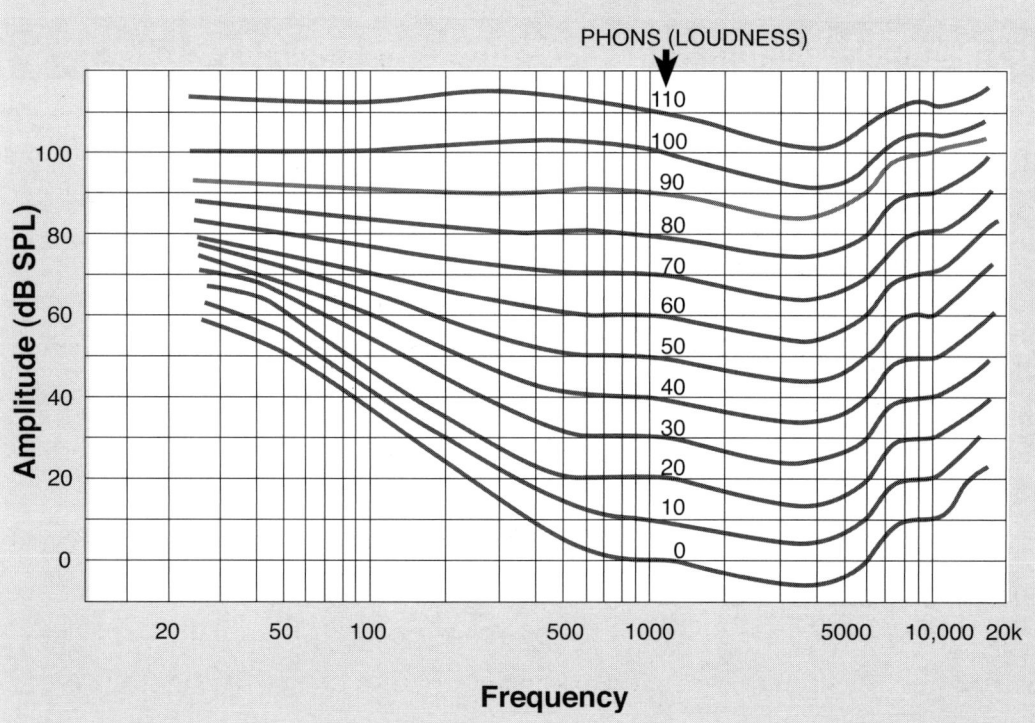

The SPL Meter

A simple and inexpensive sound pressure level meter is a valuable tool. Weighting can be switched between A and C, and response time is adjustable from slow to fast. The calibrated microphone is built in at the top of the device.

A meter like this is designed to held in space and, as much as possible, kept from the influence of reflections from surrounding surfaces such as surrounding walls or the operator's body.

To use this tool effectively, stand facing 90 degrees away from the source with the SPL meter directly in front of you and at arm's length. The built-in mic is omnidirectional so the meter doesn't need to face the source.

This SPL meter is set to read 90 decibels plus or minus 10 decibels. This 20-dB range is adequate for most situations.

C-weighting is optimized for full-bandwidth sources at levels exceeding 85 dB. A-weighting filters out the high and low frequencies and is optimized for lower volumes. The A-weighted scale more closely reflects perceived volume, whereas the C-weighted scale measures amount of energy (amplitude).

Phase

We discovered previously that a sound wave is represented by one complete cycle—a crest and a trough—which is measured along the timeline in degrees. The beginning of the crest is at zero degrees and the end of the trough is 360 degrees. The way multiple sound waves interact in the same acoustical or electrical space is called phase.

Since a sound wave has a crest, which pushes on your eardrum, and a trough, which pulls on your eardrum, it's fairly simple to visualize that if two identical waveforms happen simultaneously and follow the exact same path, their energy would increase as they worked together—in fact, they double in amplitude, meaning the peak is twice as high and the trough is twice as deep. As seen by your

Microphones & Mixers... by BILL GIBSON

eardrum, the compression and rarefaction are doubled. Two identical waveforms, which start at the exact same point in time and follow the identical path through the crest and trough, are said to be in phase.

If two signals are out of phase, their waveforms are mirror images of each other. The electronic result of this combination is silence. When this happens electronically, the energies oppose each other completely—for each push there is an equal pull throughout all 360 degrees. Since we refer to a complete cycle as 360 degrees, we mark the center point of the cycle at 180 degrees. By delaying one

A-, B-, and C-Weighting

Any piece of gear that quantifies amplitude must specify whether it's sensitive to a full or limited bandwidth. Weighting is the qualifier for sound pressure level measurements.

C-weighting closely approximates full-bandwidth sensitivity. This is the scale that most accurately represents amplitude.

A-weighting closely approximates loudness, attenuating the lower frequencies to resemble the response of the human ear (which is most sensitive to frequencies between 1,000 and 4,000 Hz).

B-weighting includes more of the mid frequencies in its sensitivity than A-weighting. It's usually used in conjunction with A- and C-weighting in analysis of acoustical anomalies.

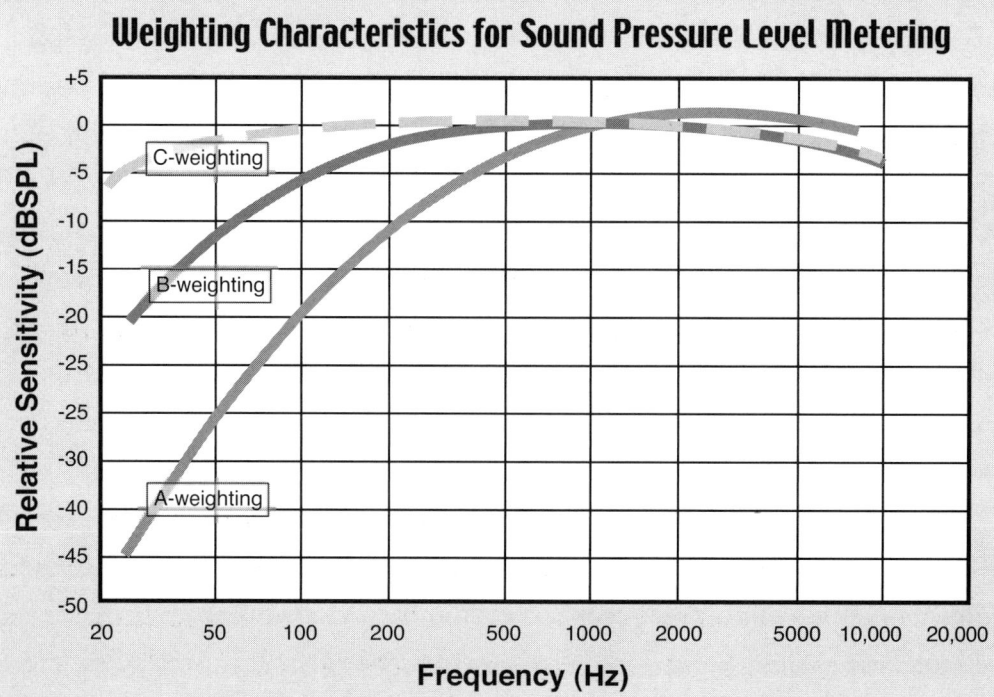

of two identical waveforms so that the beginning of the trough of one coincides with the beginning of the crest on the other (180 degrees into the cycle), we create a scenario of complete phase cancellation. When this happens, we say the two waveforms are 180 degrees out of phase.

It's easy to create a scenario, electronically, where two waveforms combine 180 degrees out of phase. It rarely happens acoustically because of the predominance and complexity of reflections, along with the fact that we hear with two ears, which already receive the same waveform at slightly different points of time. Interactions between acoustic sound waves is, however, still an important factor in understanding music and recording. For our study of acoustics, it's most enlightening to recognize the concept that multiple sounds work together to form the whole.

Waveforms can combine out of phase at any point in the cycle. If two waveforms are 90 degrees out of phase, they interact together to change the resultant sound. Even though there isn't complete phase cancellation, we still experience the result of the opposing and summing forces.

There are a few different ways that negative phase interactions provide obstacles:

- When multiple microphones are used in the same room, sounds can reach the different mics at different times and probably at different points in the cycle of the wave. They combine at the mic out of phase. That's why it's always best to use as few mics as possible on an instrument or group of instruments in the same acoustical space. Fewer mics means fewer phase problems.

- This theory also pertains to the way speakers operate. If two speakers are in phase and they both receive the identical waveform, both speaker cones move in and out at the same time. If two speakers are out of phase and if they both receive the identical waveform, one speaker cone moves in while the other speaker cone moves out. They don't work together. They fight each other, and the combined sound they produce is not reliable.

Harmonics, Overtones, and Partials

Harmonics are the parts of the instrument sound that add unique character. Without the harmonic content, each instrument would pretty much sound the same, like a simple sine wave. The only real difference would be in the characteristic attack, decay, sustain, and release of the individual instrument.

16

Microphones & Mixers .... by BILL GIBSON

Phase Relationship

Wave B is 180 degrees out of phase with Wave A. The result of opposing crests and troughs is no air movement. No air movement means no sound.

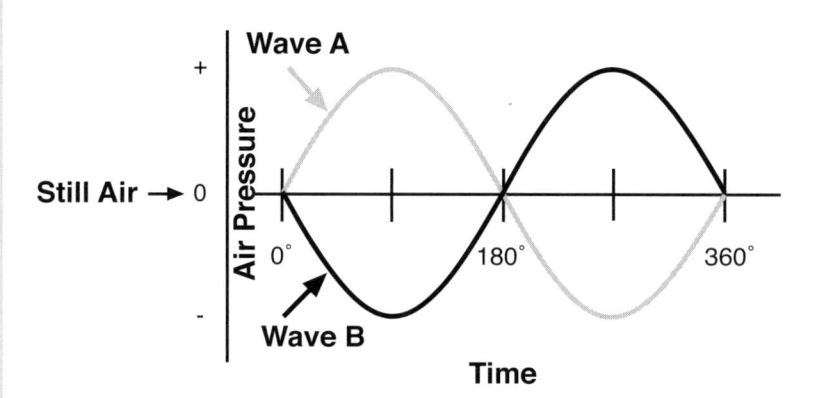

Conversely, two identical waveforms that start at exactly the same time (below) are in phase. They combine, resulting in twice as much energy.

The height of the waveform (the distance above and below the center line) is referred to as the amplitude. Amplitude corresponds to the amount of energy in the waveform. When a waveform is inverted, it is moved 180 degrees out of phase.

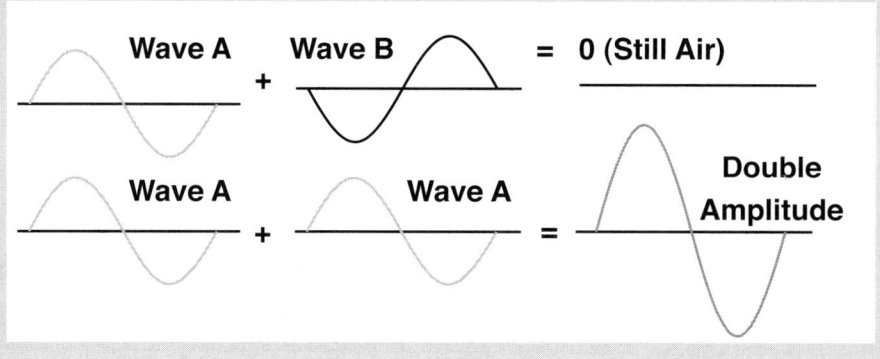

Since harmonics and overtones are so important to sonic character (vocal or instrumental), it's important to understand some basics about harmonics. As your experience level increases, this understanding will help you grasp many other aspects of music and recording.

When you hear middle C on a piano, you're hearing many different notes simultaneously that form together to make the sound of a piano. These different notes are called harmonics. Harmonics and overtones are a result of, among other considerations, vibration of the instrument; size of the instrument; acoustics; the type of material the instrument is made of; or the vibration of the string, membrane, reed, etc. Several factors add to the harmonic content, but it's a law of physics that harmonics combine with the fundamental wave to make a unique sound that is represented by one waveform. That waveform is a result of the combination of energies included in the fundamental frequency and all of the harmonics. The fundamental is the wave that defines the pitch of the sound wave.

The frequencies of the harmonics are simple to calculate. Harmonics are whole-number multiples of the fundamental frequency. In other words, if the funda-mental has a frequency of 220 Hz (A below middle C), calculate the harmonics by multiplying 220 by 1, 2, 3, 4, 5, 6, and so on.

- 220 x 1 = the fundamental, the frequency that gives the note its name, the first harmonic
- 220 x 2 = 440 Hz, the second harmonic
- 220 x 3 = 660 Hz, the third harmonic
- 220 x 4 = 880 Hz, the fourth harmonic
- 220 x 5 = 1100 Hz, the fifth harmonic
- 220 x 6 = 1320 Hz, the sixth harmonic

It's traditional to primarily consider the sonic implications of the harmonics up to about 20 kHz, since that is the typical limitation of our ears and equipment. There is a controversy regarding the importance of the upper harmonics above 20 kHz. As we understand how frequencies interact, it's not difficult to imagine that the frequencies above our audible frequency spectrum have an effect on those we can hear.

··· Video Example 1-2

Harmonics - How They Combine to Create Tonal Character

Engineers involved in archiving music and sounds for future reference carry on spirited debates about this. High-quality archival of important recordings is a big topic in the digital realm. Although digital storage seems very well suited to archiving because of its durability and long-lasting construction, the fact that CD-quality audio (at 44.1-kHz sample rate) cuts off all frequencies above 20 kHz sheds a questionable light on its long-term viability for important audio archiving.

18

Microphones & Mixers... by BILL GIBSON

Tone Interactions - Harmonics

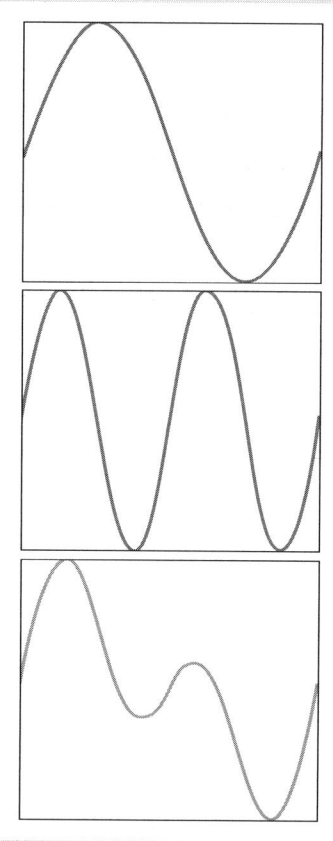

This is the fundamental sine wave. Its frequency determines the note name and pitch for the waveform that's built from it.

This is the second harmonic in relation to the fundamental above. Its frequency is two times the fundamental, so it completes its cycle twice in the same time period that the fundamental completes one cycle.

This is the result of combining the fundamental and the second harmonic. This new waveform has its own unique wave shape and sound. When waves combine, our ears no longer detect separate sound waves, they merely react to the one new wave that is influenced by all simultaneously occurring sounds.

Digital sample rates of 192 kHz or higher make more sense when considering the future of audio storage.

As the harmonics combine with the fundamental, summing and canceling occur between the fundamental and its harmonics. This summing and canceling interaction is what shapes a new and different-sounding waveform each time a new harmonic is added.

The terms harmonic and overtone are often used synonymously, but there is a difference. Whereas the harmonics are always calculated mathematically, as whole-number multiples of the fundamental, overtones are referenced to intervals and don't always precisely fit the harmonic formula. In the case of the piano, for example, the overtones are very close to the mathematical harmonics, but some are slightly off.

Some percussion sounds contain a relative of harmonics and overtones called partials. Like overtones, partials aren't mathematically related to the fundamental in the same simple formula as harmonics, and the effect that these sounds has can be very dramatic and interesting. Some bell-type sounds contain partials that are very far removed from the true harmonics (sometimes they even sound out of tune), but the overall sound still has a defined pitch with a unique tonal character. Partials can also be lower in pitch than the fundamental, whereas harmonics and overtones are considered to be above the fundamental. On bells, there's generally a strong partial at about half the frequency of the fundamental called the hum tone.

Harmonics, overtones, and partials extend far beyond the high-frequency limitations of our ears. For example, when we hear the lowest piano note, we're really hearing the fundamental plus several harmonics working together to complete the piano sound. If we only consider that the piano contains fundamental pitches from 27.50 to 4,186.01 Hz, it might not seem important to have a microphone that hears above 4,186.01 Hz. However, if we understand that for each

Wave Shapes

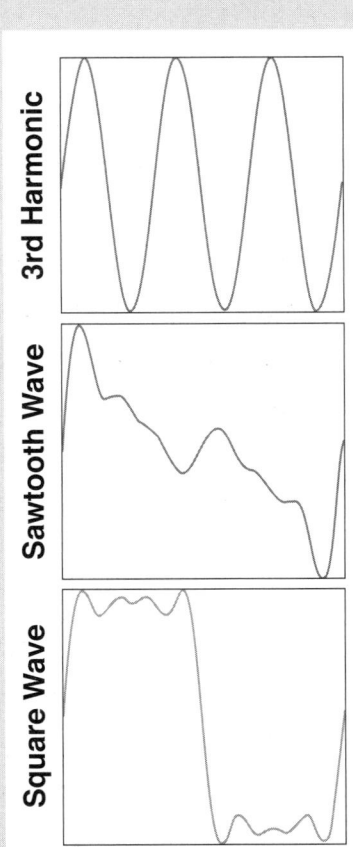

3rd Harmonic

This is the third harmonic of the sine wave in the previous illustration. Notice that each harmonic is also a sine wave, but when they're combined with the fundamental and the other harmonics, an entirely new and unique waveform is created.

Sawtooth Wave

This is a sawtooth waveform. It's created by combining all harmonics in proper proportion. The sawtooth and triangle waveforms have a bright, edgy sound. Waveforms are given descriptive names based on the shape of their sound wave.

Square Wave

This is a square wave. It's created by combining the odd harmonics (1, 3, 5, 7, 9, etc.) in the proper proportion. A square wave sounds much like a clarinet.

Microphones & Mixers .... by BILL GIBSON

fundamental there are several harmonics, overtones, or partials sounding simultaneously that go up to or above 20 kHz, we realize the importance of using equipment (mics, mixers, effects, and recorders) that accurately reproduces all of the frequencies in and/or above our hearing range. Also, if we see that the combination of these fundamentals and overtones is what shapes the individual waveform, it becomes evident that if we want to accurately record a particular waveform, we should use a microphone that hears all frequencies equally. If the mic adds to or subtracts from the frequency content of a sound, then the mic is really changing the shape of the waveform.

Shape

There are some traditional wave shapes that we refer to when describing sounds. Sine, sawtooth, square, and triangle waves each have distinct characteristic sounds. Sounding a lot like a flute, the sine wave has the simplest shape.

A sine wave is a smooth and continuous variation in energy throughout the wave cycle, steadily increasing in compression, then gently cresting over the peak to fall smoothly into rarefaction, then rising gently back to the center line. This simple waveform is also called a pure tone.

The fundamental is a sine wave, and each of the harmonics and overtones is also a sine wave. When the fundamental combines with its harmonics to create a new and unique waveform, they create a complex waveform.

It's important to realize that when we hear the fundamental and its harmonics, overtones, or partials, we don't hear any of the individual sine waves. Instead, we hear the result of the combination of all waves as one distinct waveform. The relative level of the harmonics determines their effect on the fundamental frequency, therefore shaping and molding the waveform and creating a sonic character. This unique blend of the fundamental frequency and its harmonics is called tone quality, color, or timbre.

Sawtooth, square, and triangle waves get their names from the overall shape of their unique wave. Sawtooth and triangle waves are edgy sounding and have more of a brass and bright string-type sound. A square wave sounds like a clarinet.

The complexity of the piano waveform is the result of a rich harmonic content. Piano is an instrument full of interesting harmonics. Listen carefully to a low note on the piano and notice the complexity of the sound of a single note. If you listen closely, you can isolate and hear several pitches occurring with the

Chapter 1 .... SOUND THEORY

21

The Piano Waveform

This is the actual waveform of a single piano note recorded in stereo. The top waveform is the signal from the mic placed over the low strings. The wave on the bottom is the signal from the mic placed over the high strings. Notice the complexity of these waveforms compared to the sine waves in the previous illustration.

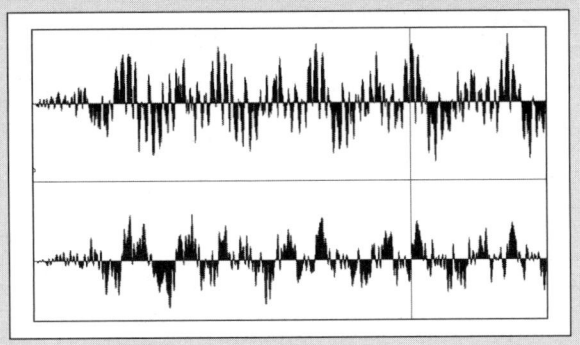

fundamental. We perceive the harmonic content, along with the fundamental, as one sound. In actuality the single piano note is constructed of many sine waves combining to give the impression of a single note with a unique timbre.

The illustration on this page has the fundamental wave drawn on top of the piano sound wave. This fundamental wave is very simple, yet the sound of the piano is very complex.

If you understand the theory of harmonics, you're well on your way to understanding the theory of sound. You'll also approach music and sound with a little more respect, finesse, and insight.

Envelope

The envelope describes the initial action, development, and diminishing of a waveform over the course of time. There are four primary phases of the envelope of any waveform: attack, decay, sustain, and release.

Attack

The way a sound is initiated is called its attack. The attack phase sees the amplitude rise from zero to the attack peak. Sometimes the attack rises to exactly the level of the sustain phase; other times the attack contains a peak attack, called a transient, which falls slightly, after the initial attack, to enter the sustain phase. Most sounds with extreme attacks contain a transient, which exceeds the average level of the overall sound.

Examples of sounds with fast attack times are:
+ Wood block

22

Microphones & Mixers.. by BILL GIBSON

The Piano Waveform

Notice the purple line drawn on top of the piano waveform. This line represents the fundamental frequency of the piano note. The fundamental frequency is really nothing more than a simple sine wave.

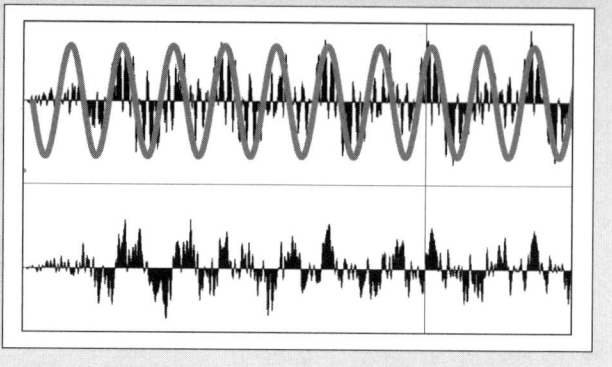

- ✦ Slap
- ✦ Snare drum
- ✦ Acoustic guitar played with a pick

Examples of sounds with slow attack times are:

- ✦ A violin, gently starting a long tone
- ✦ The swell of a Hammond B-3
- ✦ The sound of a crash cymbal reversed
- ✦ The sound of an approaching helicopter

Decay

When the peak attack is reached, the energy might decrease quickly following the peak, or it could diminish slowly until it reaches constant amplitude. The reaction of the amplitude after the attack is called decay.

Sustain

Once the sound has leveled from the attack and the decay, the period that the sound is still generating from the source is called sustain. Sustain is dependent on the generation of sound at the source. As long as the source continues, the waveform is sustaining. The sustain phase can remain at a constant amplitude, increase in amplitude, or decrease in amplitude.

Release

Once the source stops generating the sound, the envelope enters the release phase. The easiest example to explain the release phase is reverberation (natural or simu-

The Envelope—ADSR

How sound develops, holds, and decays over time comprises the envelope. The envelope parameters that we use to describe sound's amplitude characteristics over time are attack, decay, sustain, and release.

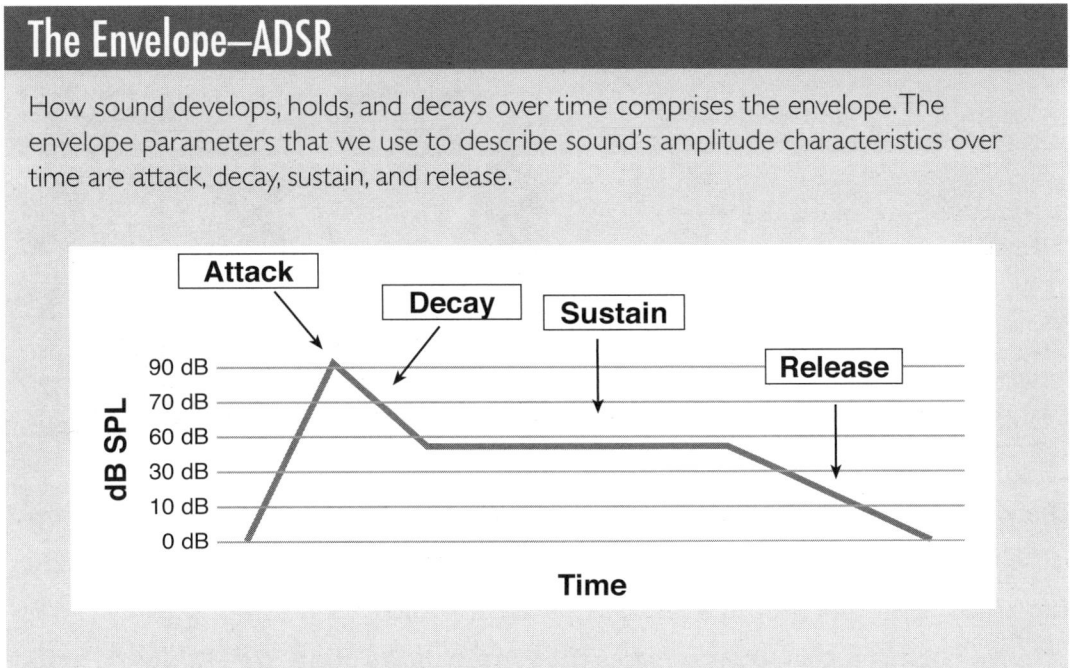

lated). When a violinist stops and removes the bow after a long, sustained note, the sound of the violin fading away in the concert hall represents the release.

HEARING PROTECTION

As a recording enthusiast and professional, your ears are your primary tools—protect them! Our awareness of the damaging effects of prolonged wear and tear has increased over the years. If you subject yourself to loud sounds and music for long periods of time, you will most probably damage your hearing.

Hearing damage is typically manifested as a decrease in your ability to hear high frequencies and/or as a persistent and annoying ringing tone that is present whether or not you're in the presence of sound.

Modern hearing protection devices are very effective and should be worn whenever you are in the presence of loud sounds. Many devices are designed to simply close off the ear canal to one degree or another. They might affect all frequencies, but they typically affect the high frequencies the most.

Some manufacturers provide earplugs that simply reduce the gain by a specified number of decibels while still letting the user hear a full-bandwidth sound—it's as if they just turned down the sound by 15 to 30 or 40 decibels. I use a set of these whenever I go to hear a band that wants me to produce them or whenever

24

Microphones & Mixers..by BILL GIBSON

Molded Ear Pieces

These molded ear pieces fit snugly in the ear canal, closing off the ear to outside sounds and reducing the broadband sound source in level by 15 dB.

the volume is excessive and out of my control. I don't use them during mixdown or when I happen to mix a live show but I do use them whenever possible—they are an excellent tool for the audio enthusiast or professional and they provide a listening experience that approximates the tonal balance of natural unfiltered audio.

When you are mixing, tracking, or just listening to music, you can control the volume. Avoid the tendency to listen at extreme volumes. Part of the reason some of us like to listen to music at potentially dangerous volumes is because our ears act as natural compressors—the mix starts to sound more blended and powerful at high levels, plus the actual movement of the air provides a hypnotic and almost addictive sensory stimulation. When music can be felt as well as heard, it provides a heightened experience.

Don't be fooled into thinking that the only damaging sounds come from a rock and roll band. Symphonic orchestras can easily approach or surpass the volume of many contemporary bands. An aggressive Big Band performing all out is a volume force to be reckoned with.

EQ Versus Volume

Very often, someone will complain about the mix being too loud, but what they're really hearing is a particular frequency that is out of balance. Vocals that are piercing typically contain an abundance of frequencies between 2 and 3 kHz. To many people, especially the elderly, those frequencies are painful. For them, the mix might sound too loud even when the decibel meter reads in the acceptable range.

Once you understand the proper use and application of equalization, you'll be able to construct a mix that is full and pleasing to the majority of listeners, even at a full and powerful level.

Whereas the highs are often too edgy and harsh for some people, low frequencies can be tolerated by most. Try to create a mix that is full in the low end and inoffensive in the highs. This type of mix will typically receive the most accolades from the largest segment of the audience.

Listen to the mixes in Audio Example 1-1. They are all the same peak amplitude, but notice how different they sound in volume.

. . . . . Audio Example 1-1

The Effect of EQ on Perceived Loudness—Mix Peaked at 2, 3, 4 kHz Plus 100, 200, 50 Hz

Painful Frequencies Versus Comforting Frequencies

The previous examples reveal that even mixes created at identical signal levels (readings on the output meter) can sound louder or softer depending on the frequency balance across the audible spectrum. Typically, elevated frequency levels between 1 and 4 kHz seem louder than elevated frequency levels below 1 kHz or above 4 kHz.

As noted earlier in this chapter, the human ear does not have a flat frequency response across the frequency or volume spectrum. It is most sensitive to the frequency band between 1 and 4 kHz, especially at volumes below 80 dB SPL.

Low frequencies, below about 150 Hz, tend to round out the sound—they help the audio engineer create a full-sounding mix. A mix with clean highs above 7 or 8 kHz and warm lows below about 150 Hz typically sounds good at reasonable levels—it usually doesn't need to be loud to sound full. As we work our way through equalization and other sound-shaping techniques, we'll discover how to balance the frequency spectrum throughout the mix in a way that sounds good and is non-offensive.

Understanding Hearing Differences

Aurally speaking, we are not all created equally. Some people are quantifiably more sensitive to certain sounds than others—it's just the way they are made.

Many musicians and music lovers enjoy the sensation of loud music. As the sound gets a little louder (90 dB and above), the low frequencies can be felt as well as heard, and the highs are clearer and more precise. Loud volume isn't perceived as anything other than a pleasurable listening experience. When someone complains about loud volume or about loud sounds causing pain, it's difficult for these

Microphones & Mixers... by BILL GIBSON

lovers of loud music to relate to that person's complaint or empathize with his or her pain.

Wise audio engineers will become educated about hearing issues in order to conserve their own hearing as well as to understand the hearing of those they work with and record.

Types of Hearing Loss

There are three basic types of hearing loss (presbycusis, tinnitus, and conductive) and two fundamental physiological conditions that cause hypersensitivity to loud sounds (hyperacusis and recruitment).

Hearing loss can be caused by loud sounds, age, and obstructions, such as swelling and wax buildup. Each person seems to experience a unique set of symptoms in this area. When subjected to the same sound source, some experience little damage, while others experience catastrophic damage.

Presbycusis

Presbycusis is hearing loss that is primarily related to aging; however, illness, prescription drugs, circulation problems, loud noises, heredity, infection, or head injury can also cause it. This loss occurs gradually over time and is typically due to sensorineural damage, in which there is damage to parts of the inner ear, the auditory nerve, or hearing pathways in the brain. The progression of presbycusis is often analogized to the body's transformation of one's original hair color to gray—it is gradual, steady, and relentless, and it happens at differing times and speeds for everyone.

People with presbycusis usually experience the regular discomfort of hearing loss—difficulty understanding speech and other sounds that contain abundant high-frequency information. In addition, this condition can cause intolerance of loud sounds.

Tinnitus

Tinnitus manifests itself as a ringing, roaring, or other noise that's heard separately from acoustic sounds. Although tinnitus accompanies other forms of hearing impairment, it also adds to hearing loss in varying degrees, depending on the severity of the internal noise.

Chapter 1 ....SOUND THEORY

27

The noise of tinnitus is typically constant, although it varies in intensity, depending on several factors. It can be caused by loud noises, hearing loss, medication, jaw misalignment, circulatory problems, certain tumors, allergies, ear or sinus infections, wax buildup, and head or neck trauma.

People who attend loud concerts often experience a ringing in their ears after a show—this is tinnitus. Sometimes the ringing will disappear over time, but hearing damage is cumulative. Continued concert attendance could result in an increase in the length of time the ringing persists. Eventually, the ringing might remain permanently.

Some aids claim to decrease the symptoms of tinnitus, but most in the medical community consider this condition to be incurable as of yet. Some herbal remedies claim to help improve the condition, and vitamins C and B-12 are said, by some, to help decrease the ringing noise. Some say that aspirin, alcohol, and smoking worsen the condition.

Conductive Hearing Loss

The tympanic membrane, also called the eardrum, receives the sound waves—it vibrates sympathetically with the sounds to which it is subjected. Next, the sound information moves to the inner ear. Conductive hearing loss results when the sound information is blocked from the inner ear.

This type of hearing loss can be caused by abnormal bone growth, infection, earwax buildup, fluid in the middle ear, or a punctured eardrum.

Aural Hypersensitivity

Certain individuals are more sensitive to sounds than others. To the hypersensitive listener, even normal sounds can cause pain—loud sounds might be intolerable. There are two primary types of hypersensitivity to sound: hyperacusis and recruitment.

Hyperacusis

People with hyperacusis have essentially no hearing loss. In fact, they are abnormally sensitive to certain sounds—even common sounds cause pain or discomfort. Those who experience this condition often rely on earplugs and other noise reducers to participate in everyday activities.

Microphones & Mixers .... by BILL GIBSON

Occasionally, tinnitus accompanies hyperacusis. The potential causes for this hypersensitivity to sounds are suspected to be noise exposure causing subtle damage to the ear, head injuries, and dysfunctions in brain chemistry.

Many people with hyperacusis rely on earplugs to shield their ears from common everyday sounds, even at low volumes. It has been discovered that this practice increases the problem rather than aiding in it. Sufferers of hyperacusis are encouraged to use ear protection only when they are in the presence of potentially damaging sounds above 85 dB.

Recruitment

Recruitment refers to abnormal loudness sensitivity, which often accompanies sensorineural hearing damage. A person with recruitment differs from a person with hyperacusis in a couple primary ways:

- Someone with recruitment is primarily bothered by loud sounds, whereas the person with hyperacusis also experiences discomfort with moderately loud sounds.
- Recruitment is present in persons with hearing loss—those with hyperacusis do not have hearing loss.

A person with recruitment might simply feel bothered by loud sounds, but in acute cases this condition is very uncomfortable. In addition, the difference between acceptable and unacceptable volume might only be a matter of a few decibels—sound pressure levels don't necessarily need to increase dramatically to surpass the threshold of comfort.

When someone complains about volume, he or she might be suffering from recruitment and not know it. That person might sincerely believe your mix is way too loud—that it hurts—and you might sincerely believe that your mix is perfectly acceptable in volume, tone, and balance. Each of you is correct from your own perspective.

Earplugs are not always helpful in this instance—there is already hearing loss present, and it is probably most extreme in the high frequencies. Simply using foam earplugs might minimize the high frequencies that are necessary for speech discernment, while negligibly affecting the frequencies that cause discomfort.

A possible solution for the recruitment sufferer is the modern hearing aid. Digital units are capable of adjusting for many deficiencies in hearing, including overall damage, specific frequency deficiencies, and recruitment issues.

Education Is the Answer

Once you understand some of the causes behind comments that you receive during the course of your audio career, you're likely to find it easier to sympathize with the volume-sensitive crowd. They're not crazy, and neither are you.

When you are educated about potential hearing differences and the effects of loud sounds and hearing damage, you'll be more likely to protect your own hearing as well as the hearing of those with whom you work and play.

Microphones & Mixers.. by BILL GIBSON

CHAPTER TEST

1.   The simplest of sound waves is the _____ wave, and its sound is similar to a _____.
     a.   square, flute
     b.   sine, trumpet
     c.   sine, flute
     d.   square, trumpet

2.   Sound interacts with the ear by:
     a.   sympathetic vibration
     b   pushing and pulling
     c.   compression and rarefaction
     d.   b and c
     e.   All of the above

3.   Sound always travels at a rate of 1,126 feet per second.
     a.   True
     b.   False

4.   In order for two waveforms to be completely out of phase, one must be offset from the other by _____. Electronically combining these two waveforms results in _____.
     a.   180 degrees, cancellation
     b.   360 degrees, twice the amplitude
     c.   90 degrees, cancellation
     d.   180 degrees, twice the amplitude

5.   The G above middle C on the piano has a frequency of _____ and a length of _____.
     a.   392 Hertz, 2.86 feet
     b.   784 Hertz, 1.43 feet
     c.   a and b
     d.   None of the above

6.   The typical frequency response range of the human ear is:
     a.   The same as the piano
     b.   27 Hz to 4,186 Hz
     c.   20 Hz to 20 kHz
     d.   a and b

7.   As indicated by the _____, the human ear is most sensitive _____.
     a.   phase curve, when it's in phase
     b.   Ear Institute, at volumes in excess of 120 dB
     c.   Munson Phase, below 1 kHz
     d.   Fletcher-Munson Curve, between 1 and 4 kHz

8.   According to OSHA regulations, it's okay for a recording engineer to listen to music at 100 dB for up to eight hours per day.
     a.   True
     b.   False

9.   When using a decibel meter, _____ most closely approximates loudness as perceived by the human ear.
     a.   the Fletcher-Munson Curve
     b.   the phase relationship
     c.   A-weighting
     d.   transient instruments

10. The result of combining two identical waveforms in phase is:
     a.   cancellation
     b.   twice the loudness
     c.   no sound
     d.   twice the amplitude
     e.   b and d

11. The result of combining a waveform with an identical waveforms that's 180 degrees out of phase is:
    a. cancellation
    b. twice the loudness
    c. no sound
    d. twice the amplitude
    e. b and d

12. In Hertz, the first five harmonics of a 200-Hz tone are:
    a. 300, 400, 500, 600, 700
    b. 300, 500, 700, 900, 1,100
    c. 400, 600, 800, 1,000, 1,200
    d. 400, 800, 1,600, 3,200, 6,400

13. The wave that defines the exact pitch of a sound is called the:
    a. key
    b. tonic
    c. root
    d. fundamental

14. Removing all of the _____ from a timbre leaves a _____ that sounds much like a flute.
    a. sparkle, dull tone
    b. weighting, tonal curve
    c. harmonics, pure tone
    d. overtones, velocity

15. The envelope consists of:
    a. attack, delay, suspension, ritardando
    b. intro, verse, chorus, bridge
    c. the flap and the pocket
    d. attack, decay, sustain, release

16. A very fast attack, such as that from a percussive instrument, is called a _____.
    a. chiff
    b. transient
    c. sibilance
    d. phantom

17. A constant ringing in the ears, even in the absence of actual sound, is called _____.
    a. tinnitus
    b. ringing audiology
    c. rarefaction
    d. ring tone

18. There are two fundamental physiological conditions that cause hypersensitivity to loud sounds. They are:
    a. aural and binaural
    b. hypertensive and appoggiatura
    c. transformation and weighting
    d. hyperacusis and recruitment

19. A mix that is overly abundant in _____ can be edgy, irritating, and painful to the listener.
    a. sub bass
    b. midrange
    c. lows
    d. 3 – 4 kHz

20. Tinnitus is:
    a. a ringing sound in the ears
    b. caused by exposure to loud sounds
    c. cumulative
    d. avoidable through the use of hearing protection and limited exposure to loud sounds
    e. All of the above

Test answers are on page 293

Connecting

The way we connect each component and ingredient of the recording setup determines several critical factors: ease of use, integrity of signal, distortion characteristics, and other critical audio quality specifications. It's important to achieve a solid understanding of the way ingredients were designed to connect together, to learn what to look for to determine compatibility between devices, and to develop a setup that is intelligently designed for maximum flexibility and uncompromising signal integrity. The information in this chapter is fundamental to the understanding, development, and maintenance of an excellent setup.

CONNECTORS

We encounter several types of connectors when building a system and hooking together audio equipment. In this section, we cover RCA, 1/4-inch, Speakon, banana, XLR, stereo, and mono connectors as well as adapters, plugging in, powering up/down, grounding, and hums.

RCA Connectors

RCA phono connectors are the type found on most home stereo equipment and are physically smaller than the plug that goes into a guitar or keyboard. RCA phono connectors are among the least expensive connectors and were very common in home-

3 4

Microphones & Mixers.. by BILL GIBSON

RCA Phono Connectors

RCA phono connectors are most typically used in home stereo configurations. They're small, inexpensive, and not usually used in a professional application other than for making S/PDIF digital connections. With one contact for the hot lead and one for the shield, this connector is used in unbalanced applications.

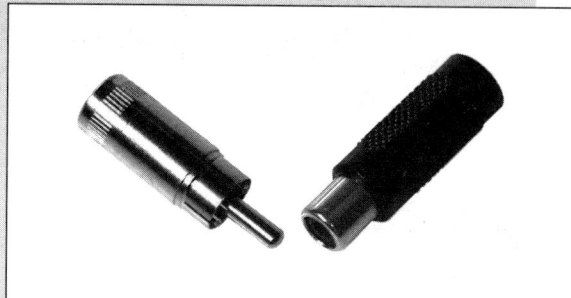

recording equipment manufactured in the mid '80s to the mid '90s. By today's standards, though, they are seldom used for serious audio connection. They are only appropriate for unbalanced applications and are virtually always used for high-impedance connections, such as those from consumer CD and DVD play-

1/4-inch Phone Plug Tip-Sleeve (Mono/Unbalanced)

The 1/4-inch phone plug is most commonly used in an instrument cable. A regular guitar or keyboard is connected to an amplifier or mixer line input with a cable utilizing a 1/4-inch phone connector at each end. This connector carries an unbalanced mono signal. Additionally, phone plugs are often used for speaker cables, using speaker wire rather than line-level instrument cable.

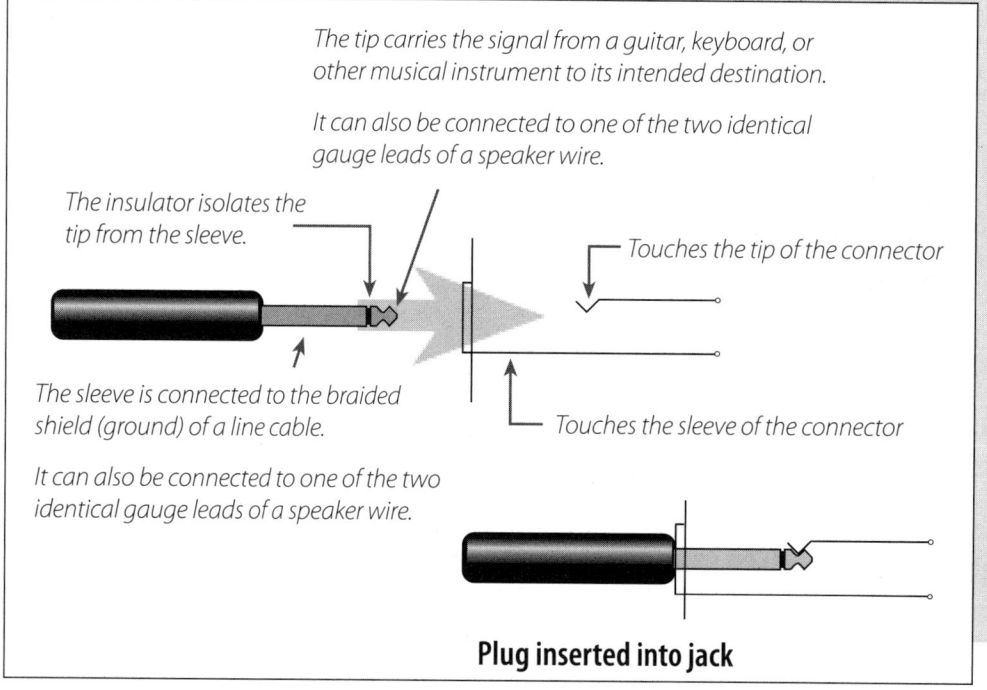

The tip carries the signal from a guitar, keyboard, or other musical instrument to its intended destination.

It can also be connected to one of the two identical gauge leads of a speaker wire.

The insulator isolates the tip from the sleeve.

Touches the tip of the connector

The sleeve is connected to the braided shield (ground) of a line cable.

Touches the sleeve of the connector

It can also be connected to one of the two identical gauge leads of a speaker wire.

Plug inserted into jack

ers, some computers, and other music players. RCA connectors are also used to make S/PDIF digital connections.

Quarter-Inch Phone Connectors

Quarter-inch phone connectors are the type found on regular cables for guitars or keyboards. These connectors are commonly used on musical instruments and in live sound systems, at home and in professional recording studios.

Notice that a guitar cable has one tip and one sleeve on the connector. In a guitar cable, the wire connected to the tip carries the actual musical signal. The wire carrying the signal is called the hot wire or hot lead. The sleeve is connected to the braided shield that's around the hot wire. The purpose of the shield is to diffuse outside interference, such as electrostatic interference and extraneous radio signals.

1/4" Phone Plug Tip-Ring-Sleeve (Stereo/Balanced)

The 1/4" tip-ring-sleeve phone plug is most commonly seen on stereo headphones. In this application, the tip and ring connections carry the left and right channels of a stereo headphone send. This connector, like the XLR, is also commonly used to carry balanced signals, where the tip and sleeve carry the audio signal like pins 2 and 3 of the XLR connector.

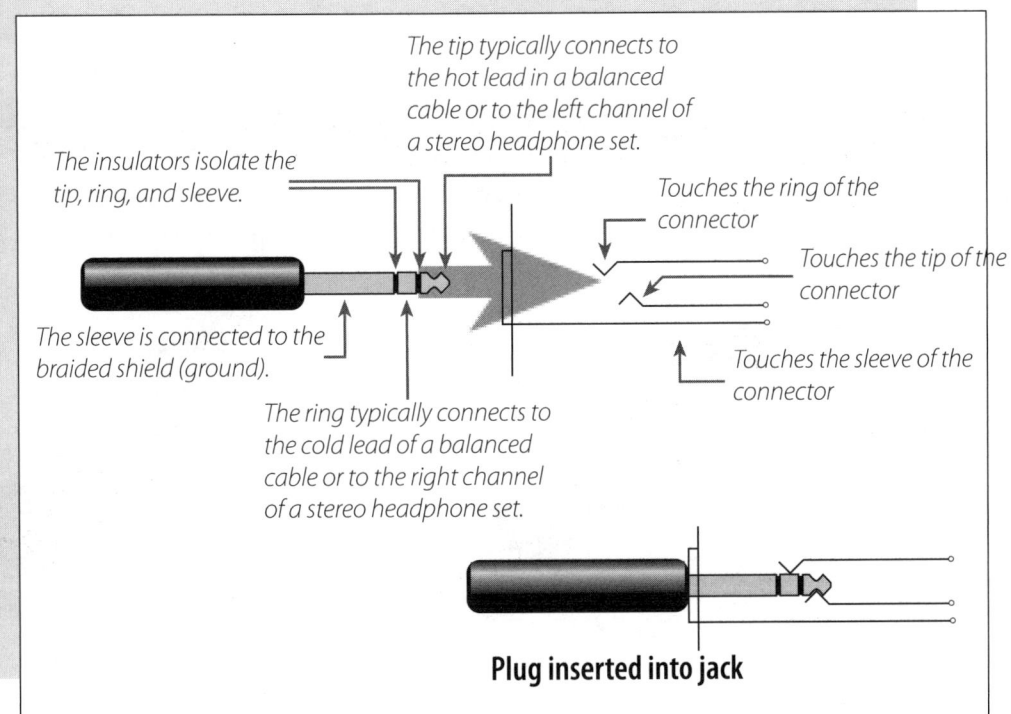

36

Microphones & Mixers .... by BILL GIBSON

The other type of 1/4-inch phone connector is the type found on stereo headphones. This plug has one tip, one small ring (next to the tip), and a sleeve. These connectors are referred to as 1/4-inch TRS (tip-ring-sleeve). In headphones, the tip and ring are for the left and right musical signals, and the sleeve is connected to the braided shield that surrounds the two hot wires. The 1/4-inch TRS connector is also commonly used for balanced line-level connections. This connector can be used for other devices that require a three-point connection.

XLR Connectors

XLR connectors are the type found on most microphones and the mic inputs of most mixers. Two of the three pins on this connector carry the signal, and the third is connected to the shield. A cable that uses XLR connectors typically carries a balanced signal, utilizing a hot lead (audio signal in positive polarity), a cold lead (identical audio signal in negative polarity), and a shield.

It's not uncommon to find cables with an XLR on one end and a 1/4-inch phone plug on the other, or cables that have been intentionally wired in a nonstandard way. These are usually for specific applications and can be useful in certain situations. Check wiring details in your equipment manuals to see whether these will work for you.

There are other types of connectors, but RCA phono, 1/4-inch phone, and XLR are the most common. It's okay to use adapters to go from one type of connector to another, but always be sure to use connectors and adapters with the

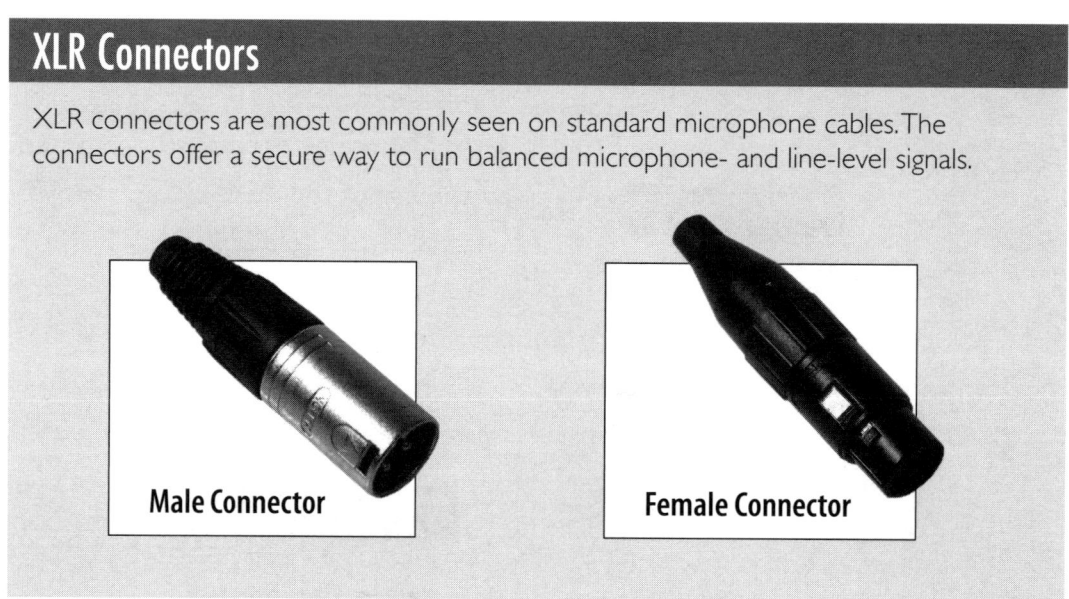

XLR Connectors

XLR connectors are most commonly seen on standard microphone cables. The connectors offer a secure way to run balanced microphone- and line-level signals.

Male Connector

Female Connector

Dual Banana Connectors

Dual banana connectors are typically used on speaker cables, often connecting at the amplifier output and at the speaker enclosure input. This connector is quick and easy to use and to assemble. There is a tab on the negative pole to verify polarity.

Dual Banana Connector

same number of points. For example, if a plug has a tip, ring, and sleeve, it must be plugged into a jack that accepts all three points in order to maintain a consistent function. In some cases, a balanced source (XLR or TRS) delivers an unbalanced signal to an input. This is accomplished by simply connecting one hot lead at the input.

The Dual Banana Connector

Dual banana connectors are very common speaker wire connectors, either at the amplifier output or the speaker box input. They are quickly connected and simple to use. One side has a tab to help keep track of consistent phasing between speaker boxes. When phase reversal is necessary, they can be flipped upside down. They're also available in various colors, so color coordination of frequency splits or amplifier runs is easy. This connector is not used for line-level connections.

Speakon Connectors

Speakon connectors are used for speaker wire termination. Prior to common adoption of the Speakon connector, the dual banana connector was most common. European regulatory requirements outlawed the use of the dual banana connector, forcing the user to terminate with spade lugs or bare wire ends. This prompted the use of the Speakon connection, which has become a very common speaker box termination. This connector typically utilizes a four-point connection, letting the user simply utilize two of the points for a standard speaker connection or all four points to pass both high and low frequencies of a biamplification scheme down the same speaker cable.

38

Microphones & Mixers.. by BILL GIBSON

Speakon Connectors

These quick and convenient connectors are typically used for speaker enclosure connections. They offer a multiple-pin (4 or 8 points) configuration, which provides a simple way to connect to cabinets with bi- and triamplified requirements. Each band is run through the same multi-pair cable.

Speakon Connector

These connectors are also available in an eight-point version, providing numerous options for multiple-amplification sends. Speakon connectors offer a

Adapters

The center post on the RCA phono plug corresponds to the tip on the RCA-to-1/4-inch phone plug adapter.

The tip and the ring on the 1/4-inch tip-ring-sleeve phone plug correspond to pins 2 and 3 on the 1/4-inch phone plug-to-XLR adapter. The sleeve corresponds to pin 1. The pin numbers of the XLR connector are imprinted on the connector end itself. They're located next to the base of the pins on the male XLR connector and next to the holes on the female XLR connector.

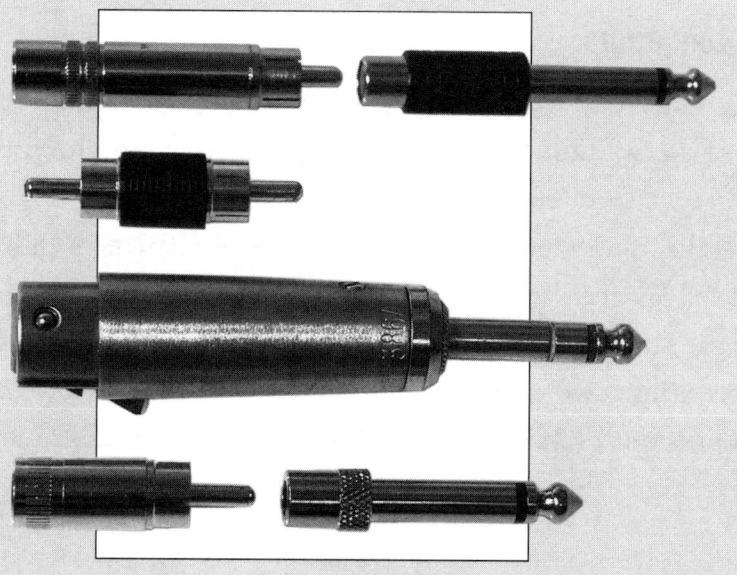

secure, twist-lock connection; they're fast, easy, and cost-effective. They are not used for line-level sends.

INTERCONNECT

When it comes to connecting and designing audio systems, wire and connectors are very important factors. There are several aspects of wire type and cable configuration that dramatically affect sound quality. It's important that we understand the physical differences between basic cable and connector types; it's also important that we experience and realize the differences between well-designed and subpar audio cables. The cable industry is full of controversy, with an ongoing battle between manufacturers who make claims of sonic superiority based on marketing claims and those who dispute those claims. I have performed several listening tests comparing cables of various cost and design and in many of those tests, I can verify that there are sonic differences—in some, I can't. Each of us should feel compelled to perform our own tests and comparisons, especially when there is a dramatic cost difference. The home hi-fi audio industry isn't bashful about charging exorbitant prices for audio and AC cables. Are they worth it? That's a personal decision we each need to make. Keep in mind, though, that underpriced cabling gets no respect in the hi-fi industry—in many cases, the higher they price it, the better it sells.

Frequently, there are considerations other than audio quality involved in cable choice, even once the appropriate size and configuration has been determined. Some people just appreciate (and can afford) buying the best of the best—it's part of the fun they have in life. Others want to provide the best tools at each point in the signal path and they assume that several small advantages will add up to a perceptible difference in the final product. While both of these traits are understandable, it's important that we make decisions based on fact and our own perception. As functional sound operators and engineers, it's important that we practice listening analytically, assessing perceived audio quality honestly and critically. On the other hand, there are some criteria that must be met based on hard facts. These include considerations regarding resistance, inductance, and capacitance.

40

Microphones & Mixers .... by BILL GIBSON

Resistance

An ohm (named after German physicist Georg Simon Ohm, who formulated Ohm's law) is a unit of resistance to the flow of alternating electrical current. A speaker or network of speakers provides resistance, measured in ohms, to the amplifier output; cable also exhibits resistance, also measured in ohms.

Typical speaker cable lengths add negligibly to the combined impedance load imposed by the speakers and cable. However, excessively long cable runs utilizing very thin cable can be problematic. Resistance throughout the cable length is cumulative with distance. Using identical wire, long runs exhibit greater impedance loads than short runs. Also, thicker cable provides less resistance than thinner cable.

It is generally understood that an addition of 1 ohm to the speaker/cable load should result in about a 0.1 dB change in level. By definition, 1 dB is the smallest incremental variance that should be perceptible to the human ear, although some listeners claim to perceive level changes of less than 1 dB.

From an article by Gene DellaSala (GDS), published on eCoustics.com, November 3, 2003:

> "The basic purpose of a cable is to transfer the signal from point A to point B unadulterated. At audio frequencies the goal is to minimize losses by controlling the amount of resistance, inductance, and capacitance. For speaker cables, we have found the primary concerns for optimal signal transfer is to minimize resistance, followed by inductance, while also keeping capacitance in check to eliminate the possibilities of amplifier oscillation or frequency peaking. For line-level analog interconnects, it's a good idea to use cables that are low in capacitance and are well-shielded to eliminate interference and external noise sources from mitigating into the signal."

Inductance and Capacitance

Inductance, measured in henrys, describes the capability of a signal path to create magnetism. Capacitance, measured in farads, describes the capability of a circuit to store an electrical charge. In the audio frequency band, the influence of inductance and capacitance is negligible.

Chapter 2 .... CONNECTING

41

From an article by Dr. Eric Bogatin, published in Printed Circuit Design & Fab, July 1, 2007:

> "Capacitance is a measure of the capacity of a pair of conductors to store charge at the price of the voltage between the conductors. The higher the capacity to store charge, the more charge can be stored for the same voltage between the conductors. Capacitance is a measure of the efficiency of storing charge, at the price of voltage.
>
> The efficiency of storing a charge does not depend on the absolute amount of charge that is currently on the conductors; it is just about the geometry of the conductors and the material properties. Bring two conductors closer, and the amount of charge they can store per volt between them increases.
>
> … Inductance is a measure of the efficiency of a conductor path to create rings of magnetic field lines, and the price you pay is the current through the conductor. You don't need current in the conductor to have an efficiency of creating rings of magnetic field lines."

LINE-, INSTRUMENT-, AND MIC-LEVEL IMPEDANCE

As stated previously, impedance is the resistance to the flow of electrical current (AC). High impedance is high resistance to the flow of electrical current; low impedance is low resistance to the flow of electrical current. If you keep that simple mental picture in mind, the rest of the details should fall into place nicely.

Terminology

- Ohm (indicated by Greek letter, omega [Ω]): An ohm is the unit of resistance to the flow of alternating current used to measure impedance.
- Impedance: Impedance is the resistance of a circuit to the flow of alternating electrical current.
- Z: Z is the abbreviation and symbol used in place of the word "impedance."
- Hi Z: Hi Z is high impedance. The exact numerical tag (in Ω) for high impedance varies, depending on whether we're dealing with input impedance or output impedance. It's generally in the range of 5,000 to 15,000 Ω for output impedance, and 50,000 Ω to 1,000,000 Ω for input impedance.

It's important here to understand that hi Z is usually greater than 5,000 to 10,000 Ω.

- Lo Z: Lo Z is low impedance. The exact numerical tag (in Ω) varies for low impedance, as well as high impedance. It's generally in the range of 50 to 300 Ω for output impedance. It's normal for microphone output impedance to be between 50 and 150 Ω, and 500 to 3,000 Ω for input impedance. Normal input impedance for lo Z mixers is 600 Ω. Essentially, lo Z usually uses small numbers below 600 Ω.

- Output impedance: Output impedance is the actual impedance (resistance to the electron flow measured in Ω) at the output of a device (microphone, amplifier, guitar, keyboard). To keep it simple, realize that the output impedance is designed to work well with specific input impedance.

- Input impedance: Input impedance is the actual impedance (resistance to the electron flow measured in Ω) at the input of a device.

Compatibility Between Hi Z and Lo Z

To keep it simple, realize that the input impedance is designed to work well with specific output impedance. Low impedance and high impedance are substantially different in their defined ranges. They are not ideally compatible.

When connecting the output of a low-impedance mic to the input of a high-impedance amplifier, there's a problem. The lo Z microphone is designed to introduce the signal to the input of a low-impedance amplifier. Typically, when a low-impedance signal meets the high-impedance input, the signal seems so weak that, even with the input level set to maximum, an acceptable level cannot be achieved. Though the connection is made and signal enters the input, the impedances of the two devices don't match, so they can't operate to their full and intended capacity.

The other incompatible scenario involves attempting to plug a high-impedance output (microphone, guitar, keyboard, and so on) into a low-impedance input (mixer, amp, speaker, and so on). In this case, the hi Z output is expecting to meet a hi Z input; in other words, the hi Z signal is expecting to meet high resistance. With the high-impedance output signal connected to a low-impedance input, the signal meets relatively little resistance and therefore easily overdrives the input—input level controls must often be set to near minimum in order to attain a level low enough for a satisfactory level.

High-impedance outputs are supposed to meet high-impedance inputs; low-impedance outputs are supposed to meet low-impedance inputs. It's not true that the input and output impedance need to be identical. In fact, the input impedance is generally supposed to be about 10 times the output impedance, but as I mentioned earlier, we need to keep in mind that high impedance uses high ratings (above 10,000 Ω), and low impedance uses low ratings (typically below 1000 Ω).

·······································Audio Example 2-1
High-Impedance Instrument into a Low-Impedance Input

·······································Audio Example 2-2
Low-Impedance Mic into a High-Impedance Input

Solution: To enable high- and low-impedance device to work together, simply use an impedance transformer—also called a line-matching transformer or direct box—to change impedance from high to low or low to high; that's the easy part. We should, however, strive to understand some of the reasons we do what we do. This simple explanation of impedance is meant to get you started toward your enlightenment. It is admittedly primary in its depth, but it functions as an excellent point of reference for further technical growth.

BALANCED VERSUS UNBALANCED

For the purposes of this course, we'll cover this topic much like we did with impedance, using simple references and, wherever possible, non-technical language. There are plenty of books about electronic circuit design and books that approach the audio world from a technical perspective; however, this book is written to help you as an operator and with that in mind, I want to make sure you understand some of the most essential basics as they pertain to helping you be a better audio engineer. Some of the differences between balanced and unbalanced wiring schemes are simple, and some are interestingly complex. As a point of reference, remember this: Almost all guitars are unbalanced and almost all microphones are balanced.

44

Microphones & Mixers...by BILL GIBSON

Terminology

+ Lead (pronounced leed): Lead is another term for wire.

+ Hot lead: In a cable, the hot lead, surrounded by a layer of insulation, is the wire carrying the desired sound or signal. From a guitar, the hot lead carries the guitar signal from the magnetic pickup to the input of the amplifier.

+ Cold lead: In a cable, the cold lead, surrounded by a layer of insulation, is the wire carrying the desired sound or signal. It is identical to the signal carried by the hot lead, although reversed in polarity. The cold lead is part of the balanced wiring scheme.

+ Braided shield: Cables for instruments, mics, and outboard gear—pretty much anything other than speaker cable—have one or two wires carrying the desired signal surrounded by very thin strands of wire braided into a tube. This braided wire, called the shield, helps diffuse, absorb, and reject electrostatic noises and interference.

Line Cables (Musical Instrument Cables)

This illustration shows the construction of typical wire used for unbalanced cables. Notice that the hot lead is stranded wire in the center core, the shield is braided wire isolated from the hot lead by a plastic tube, and around the shield is a plastic or rubber insulating material.

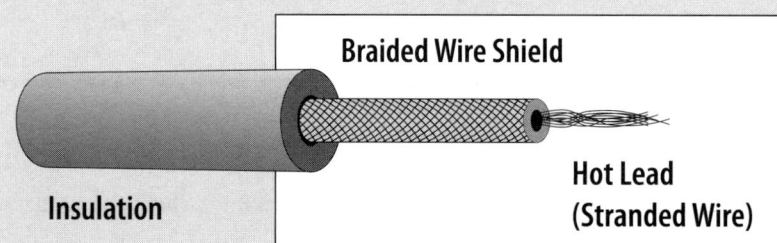

The illustration below shows the parts of a typical 1/4-inch line cable. The tip carries the actual musical signal. The sleeve is connected to the shield, which is designed to absorb, diffuse, and reject interference. The other common unbalanced connector is the RCA phono plug.

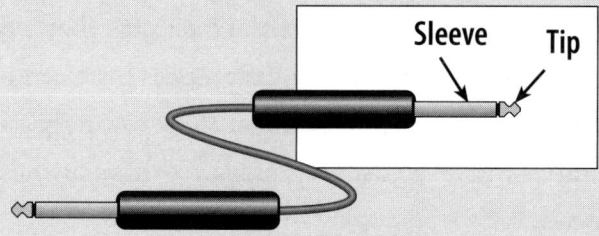

Chapter 2 .... CONNECTING

45

Unbalanced Guitar Cables

Normal guitar and keyboard cables, also called line and instrument cables, contain one hot lead to carry the instrument signal—a braided wire shield surrounds this hot lead. The purpose of the shield is to diffuse, absorb, and reject electrostatic noises and RF interference.

This system works well within its limitations. The braided shield does a pretty good job at keeping radio signals and other interferences from reaching the hot lead—as long as the cable is shorter than about 20 feet. Once the cable is longer than 20 feet, there's so much interference bombarding the shield that the hot lead starts to carry the interference along with the audio signal. The long cable acts as a crude antenna and picks up plenty of transmissions from multiple transmitters. This fact is true even when we study balanced cables; the main difference is that the balanced wiring scheme cleverly beats the system by using the system.

Balanced Wiring

Low-impedance mics, as well as most modern outboard equipment and mixers, use balanced connections. While the length limit of unbalanced cables is about 20 feet, balanced low-impedance cables can be as long as 1,000 feet, or so, without the addition of noise or electrostatic interference and without significant degradation of the audio signal.

A cable for a balanced lo Z mic uses three conductors, unlike the unbalanced system that just uses the hot lead and the braided shield. Of these three conductors, two are used to carry the signal, and the other is connected to ground. Two-conductor shielded cables are also very common. Whether two or three leads are present, they are all twisted together throughout the length of the cable so that they are exposed to the same electrostatic noises and RF interference.

The term "balanced" is derived from the fact that two leads carry the signal and that they are perfectly balanced, sharing the exact same impedance. They receive the same noise and interferences, though one is hot and the other is reversed in polarity (cold) relative to the hot lead. The hot lead is typically connected to pin 2 on the XLR connector and the cold lead, which, again, carries the identical signal as the hot lead but reversed in polarity, connects to pin 3. As a point of interest, though not crucial to our understanding of the balanced wiring concept, some European standards and some balanced line-level connections designate pin 3 as hot and pin 2 cold.

46

Microphones & Mixers.. by BILL GIBSON

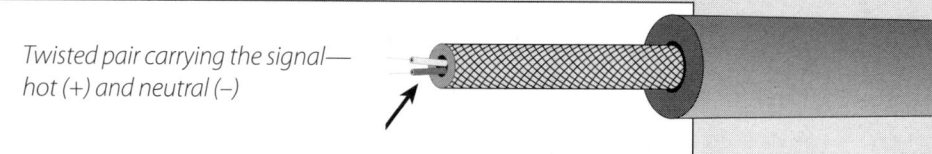

Wire for Balanced Cables (Two-Conductor Shielded)

Most wire for balanced cables has two separate leads twisted together in the center core throughout the length of the cable. Both of these leads carry the signal and connect to pins two and three; the braided shield connects to ground.

The wire connected to pin 2 is typically called the hot lead because it carries the signal in positive polarity. The wire connected to pin 3, referred to as neutral or cold, also carries the signal, although its signal is reversed in polarity relative to the signal in the hot lead.

Twisted pair carrying the signal—hot (+) and neutral (–)

Three-Point Connectors

Any three-point connector can be used on balanced cables. As long as there's a place for the two hot leads and a ground to connect, the system will work. XLR connectors are the most common, but the 1/4-inch tip-ring-sleeve plug—like the kind on your stereo headphones—is also common. In commercial studios, a smaller version of the 1/4-inch stereo plug, called the Tiny Telephone connector, is also common.

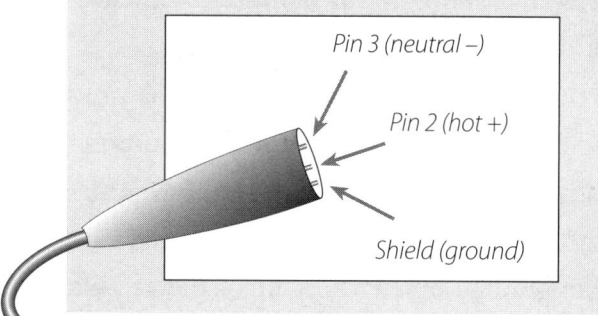

Pin 3 (neutral –)
Pin 2 (hot +)
Shield (ground)

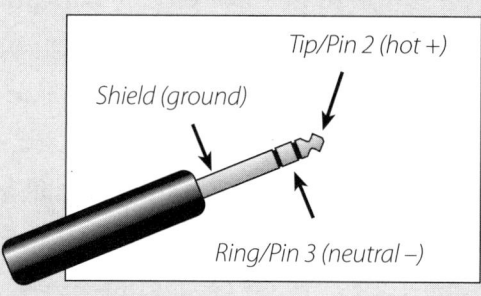

Tip/Pin 2 (hot +)
Shield (ground)
Ring/Pin 3 (neutral –)

At the source, the signal is split and sent down the hot and cold leads—the polarity of the cold lead is reversed relative to the hot lead. The hot lead is indicated by the plus sign ("+") and the cold lead is indicated by the minus sign ("–").

At the input stage (the end of the cable opposite the source), the hot and cold leads are recombined using a differential amplifier, which measures the difference in voltage between the leads. All noise and interference should be identical in the hot and cold leads and, since the polarity of the intended audio signal was reversed at the onset, it's the source audio that is different between the leads. The differ-

Balanced Wiring Theory

Pins 2 and 3 carry the identical signal; however, the polarity of pin 3 is reversed in relation to pin 2. The two wires are twisted throughout the length of the cable so they receive identical RF inference and noise. A differential amplifier at the mixer measures the difference between the two leads. Since the noise has equally influenced the twisted leads, it is equal in both and is ignored by the differential amplifier. And, since the signal is opposing in polarity between the two leads (as different as it can possibly be), it is recombined and increased in amplitude.

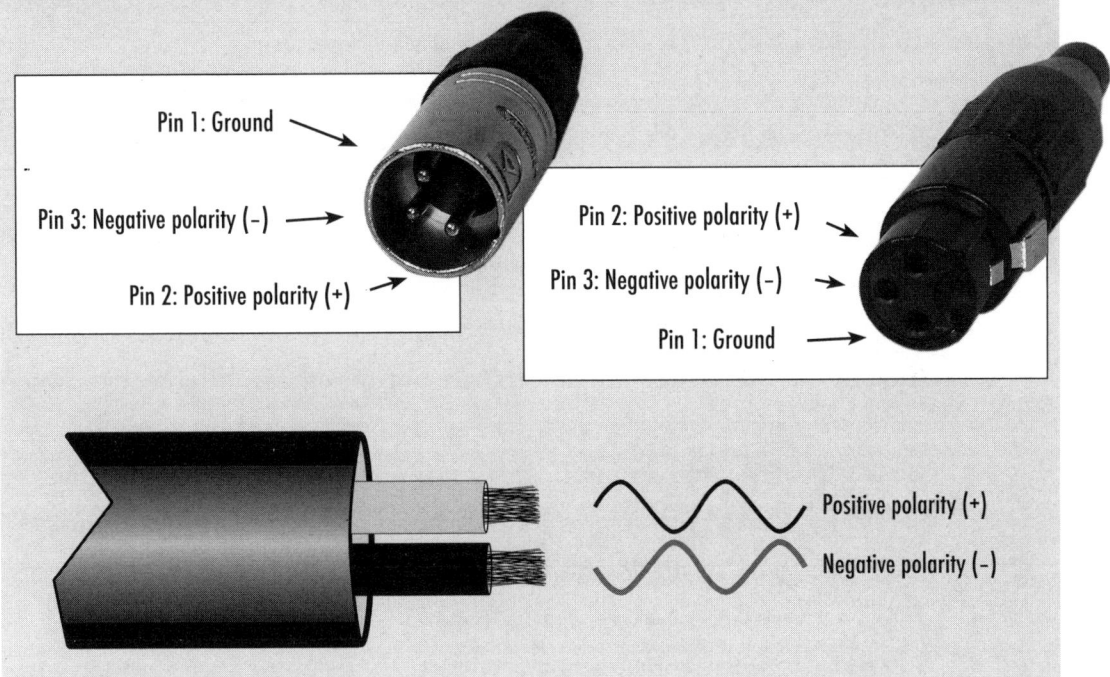

Pin 1: Ground

Pin 3: Negative polarity (–)

Pin 2: Positive polarity (+)

Pin 2: Positive polarity (+)

Pin 3: Negative polarity (–)

Pin 1: Ground

Positive polarity (+)

Negative polarity (–)

ential amplifier sees only the original audio signal as it amplifies the difference between the leads so the source audio is recombined and its amplitude is doubled. Because the electrostatic noises and RF interference have been saturating the hot and cold leads equally throughout the length of the cable, they are ignored by the differential amplifier as it recombines the hot and cold leads. Noises and interference are, therefore, eliminated.

Any three-point connector can be used on cables that connect balanced devices. As long as there's a place for the hot and cold leads and a ground to connect, the system will work. XLR connectors are the most common, but a plug such as a 1/4-inch tip-ring-sleeve configuration is also common. In large studios, a smaller version of the 1/4-inch TRS connector—the Tiny Telephone (TT) connector—is commonly used for balanced patch bay connections.

48

Microphones & Mixers .... by BILL GIBSON

In summary, the result of balanced wiring is total cancellation of noise and interference, plus a doubling in amplitude compared to the signal in an unbalanced system.

CHOOSING CABLE

The controversy has been brewing for years in the audio world. This excerpt of "A Spat Among Audiophiles Over High-End Speaker Wire," by Roy Furchgott, was taken from the December, 23, 1999 issue of the New York Times.

> "In the last year, Lewis Lipnick has tested high-end audio cables from 28 manufacturers. As a professional musician with the National Symphony Orchestra and as an audio consultant, he counts on his exacting ear to tell him if changing cables affects the accuracy of the sound from his $25,000 Krell amplifiers.
>
> His personal choice is a pair of speaker wires that cost $13,000. 'Anyone would have to have cloth ears not to tell the difference between cables,' he said.
>
> 'In my professional opinion that's baloney,' said Alan P. Kefauver, a classically trained musician and director of the Recording Arts and Sciences program at the Peabody Institute of Johns Hopkins University. 'Has the wire been cryogenically frozen? Is it flat or round? It makes no difference, unless it makes you feel better.' His choice for speaker wire? Good-quality 16-gauge zip wire."

Over the years, I've found myself straddling the chasm between these two philosophies. On one hand, I swear I can hear the differences between certain cables—I've even included examples in my books to substantiate what I'm hearing. On the other hand, I appreciate the careful and scientific approach of the objective technicians who take the stand that, "If I can't see it on my meter, it ain't different!"

My background is musical. I've spent the vast majority of my life performing, writing, arranging, recording, producing, and mixing music in the studio or at live shows. I care more about how it sounds than what it looks like on a meter. However, some of the things that I hear in the tests I've done are pretty subtle, and it's not always easy to tell if it's subtly better or worse. Some of my tests have yielded substantial audible variations with the simple swapping of one cable in the setup.

The ultimate purpose of an audio cable is simple: deliver the signal to the destination in exactly the same condition that it's received at the source. Anything other than that is simply undesirable. If a cable design is overly complex, the potential for coloration increases. To complicate the process of finding trustworthy information about cable quality and sonic integrity, our industry is rampant with adamant opinion. I've heard passionate arguments by very successful audio industry professionals that no one could hear the difference between the highest-priced speaker cable and 12/2 copper electrical wire. I've heard equally impassioned professionals proclaim the benefits of rewiring their entire systems with the most upscale cable of the day.

In my mind, it always comes down to one question: Is this going to give me even a little more accuracy? If I think there's a chance of even a slight improvement—and if I can realistically afford it—I'll take the step. I'm a proponent of the theory that several minor improvements add up to a better final product. I also don't like wasting money on silly things. I've had the discussion with some of my buddies who are ardent hi-fi audio buffs, asking them to explain the validity of spending astronomical amounts of money to build systems. I like to highlight that the music they love most was very likely tracked and mixed using Yamaha NS10s as reference monitors with something like a Crown CD300 amplifier powering them—an entire system that could be purchased for much less than the price of one of their cables.

The answer to this conversation doesn't really matter—that is, if there is an answer. Music and audio stimulate our senses and emotions. Whether at a live show, in the studio, or at home, there is a huge amount of stuff going on behind the listener's or viewer's mental curtain. Music and sound are about emotion, passion, and stimulation of our senses. It's just impossible to live in each other's skin—to hear or feel the differences imposed by this or that. Our perceptions are real to us. I'm on a quest to match reality with perception. Care to join me?

Pre-Assembled Cables

Cables aren't inexpensive; at least cables worth owning aren't, so it is not advisable to spend a lot of time searching far and wide for the least expensive, "cheapest," cables. Typically, these low-priced "bargains" are poorly constructed and fail much

50

Microphones & Mixers... by BILL GIBSON

sooner and more often than mid- or high-priced cables. I know with relative certainty that the cables I regularly purchase are likely to be functioning properly for years to come. I've also learned that the cables I grab out of the "super deal" bin are very likely to fail within the first several months, weeks, or days of their use. Therefore, in the long term these "super deals" are likely to be the most expensive cables.

For common cable lengths, it's usually less expensive to purchase pre-assembled cables than it is to purchase wire and assemble the cables yourself. Most of the major name-brand cable manufacturers provide a reliable product. The fact that they can purchase cable and connectors in bulk, and they have an efficient and inexpensive means of assembly, allows them to still make a profit while providing reliable cables, typically at a lower cost than the combined cost of the components.

Custom-Assembled Cables

Obviously, long cable runs and permanent installations requiring specific lengths of cable must be custom designed and meticulously assembled. For installations with custom cabling spanning long runs, Belden has been the most commonly used cable for a long time. They provide an excellent product and are very experienced in providing cabling for all types of systems.

Varying Levels of Soldering Technique

These two 1/4-inch connectors offer dramatically different reliability—this picture is worth 1000 words. Take great care and pride in the neatness and accuracy of your soldering tasks—spend the time it takes to get it right. You'll be much happier and you'll save a lot of heartache at the gig!

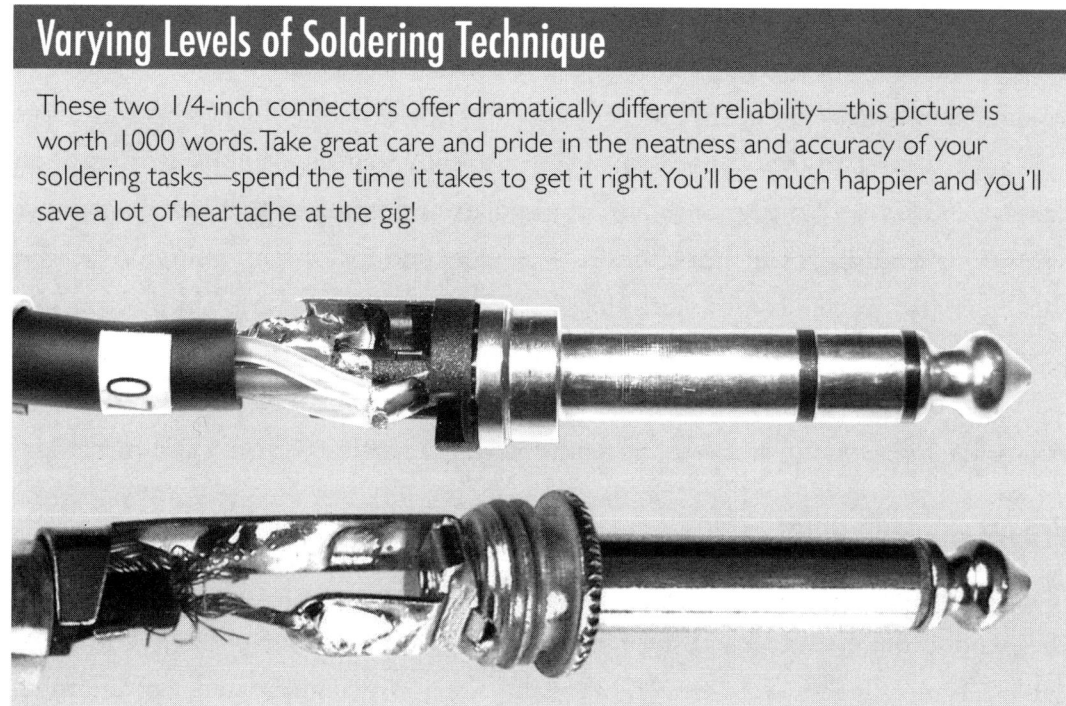

Chapter 2 .... CONNECTING

51

Assembling cables requires patience, a meticulous mindset, a steady hand, the right tools, and usually some soldering skills. There are excellent solderless connectors available that are reliable and easy to assemble. Neutrik makes excellent soldered and solderless connectors.

If you want to use cables with soldered connectors, you're uncomfortable operating a soldering iron, and you rely on the dependability of your cables, you either need to develop your soldering technique or hire an experienced technician to put everything together correctly. A connector that has been skillfully soldered onto the cable will stand up to constant use and it will last a very long time before it fails. A poorly soldered connection is most assuredly going to fail at the worst possible moment.

Probable Causes for Various Cable Failures

There are a few clues as to what's going wrong when a cable begins to fail.

1. If you hear the audio signal along with a substantial amount of noise and radio interference, it typically means that, on an unbalanced cable, the hot lead is connected but the shield is disconnected.

2. If you hear no signal, it typically means the hot lead on an instrument cable is disconnected or that both leads on a balanced connection are disconnected.

3. If you're using a balanced cable and the signal is much weaker, or quieter than normal, it typically means that either the hot or cold lead is disconnected.

It's also possible, especially in a poorly constructed cable, that two or more of the leads or shield could be touching each other inside the connector, causing noise or shorting the connection. Often, cables constructed by inexperienced technicians are on the verge of these types of failures inside the connector from the moment they are put into service.

Least Common Denominator

If you decide that you want to incorporate some highly touted and more expensive cables into your setup, do so systematically. Keep in mind that any signal chain is only as good as its weakest link. Therefore, it is senseless to have an expensive cable in the same signal chain with an inexpensive cable. When you're ready to raise the bar, the best place to incorporate new and more expensive cable is between simple connection points. Connections using just a few cables make the most sense, such as:

- ♦ Between the mixer output and the recorder input.
- ♦ Between the mixer output and the powered speaker cabinet input.
- ♦ Between both the mixer output and the power amplifiers, and between the power amplifiers and the speaker cabinets.

It makes absolutely no sense to use a $100 instrument cable to connect to an inexpensive direct box that's connected to the mixer using a $10 mic cable running through an inexpensive snake to connect to the mixer, which is connected to the power amplifier with $10 XLR cables and then connected from the power amplifier to the speaker cabinets using inexpensive lamp cord. Keep all things in perspective.

Acoustical Influences on Listening Tests

Sometimes, the room you're in exposes deficiencies in cable. I've personally experienced the effect that the room has on cable assessment. After visiting Bruce Swedien (engineer for Michael Jackson's Thriller, Bad, Off the Wall, and so on) in the early '90s, and hearing him rave about replacing all of the wire in his studio with Monster brand cable (and then after hearing music through his studio monitors), I came away convinced that cable influences the sound we hear from our systems. His studio was amazing in every way: meticulously installed, full of the best gear, huge, and sonically out of this world. Bruce is famous for being extremely influential in the development of what came to be the standard for popular commercial audio. Through his work with Quincy Jones and Michael Jackson, he led the way to powerful and pristine audio productions that remain a sonic reference standard for many excellent engineers.

After returning home, I decide to start my own search for audio purity by testing all the cables I could find. I started with the cables between the output of my console and inputs of my powered studio monitors. It made sense that this was a place in my signal path that needed to be the best it could possibly be and that it was also the least complicated checkpoint because the test involved only one cable on each of the left and right channels. The differences I heard were stark, especially in the low band. I settled on a cable constructed from Mogami quad star cable and Neutrik connectors. Especially when compared to an inexpensive (yet popular) name brand, it sounded markedly fuller in the low band and very clear in the high band. After confirming my findings with a few of my audio geek buddies, who also enjoy listening tests and shoot-outs and other such fun activities, I was pleased to implement these new cables in my studio. However, I had just moved to a new location and, when we performed the evaluation, I hadn't finished the acoustical treatment in the studio. The differences we heard were real, audible, and pronounced, but once I had finished treating the studio with diffusers, traps, and foam, we repeated the tests. What we discovered was a much less pronounced difference in the sound between the cables. Although we could still hear a similar shift in frequency content while comparing the different cables, the tonal character of the acoustical environment had been substantially tamed—the low-mid mode was much less dominant after treatment. We concluded that the effect of the various cables was real but less of a problem in my studio once it was acoustically treated.

So what's the takeaway for each of us in this battle between high-priced cable manufacturers, audio purists, technical elite, and the ardently opinionated? We should each take the time to listen and compare. We should then make our own decisions based on the data we've compiled and the experiences we've encountered. Keep in mind that cable can make a sonic difference and the environment in which we assess the cable can influence the results.

CABLE CONSTRUCTION STANDARDS

Cables from a respected manufacturer can be trusted to be reliable. Plus, we can expect these cables to deliver a reasonably accurate rendition of the signal they receive. In reality, there are very few actual cable manufacturers. Many of the cables we use are imported and the name of the apparent manufacturer is printed on the cable.

- In addition to making sure the actual wire is high-quality, it is equally important to pay close attention to the connectors used and the quality of construction. Neutrik and Switchcraft connectors have been trustworthy for years. As I mentioned previously, if you are building your own cables, be meticulous. There are standards for cable construction which, when adhered to, produce cables that can be trusted to work for a long time.

At a local AES meeting here in Seattle, Steve Turnidge (circuit designer, mastering engineer, and author), along with Aaron Gates (system designer and installer), presented the standards to which they adhere when they build cables. Steve designs circuits for Krell, Rane, and his own company, Synthwerks. Aaron has managed and implemented many full-scale installations, including Microsoft Game Studios, MSNBC Studios, and the Experience Music Project.

Their reference is the NASA wiring standard. It's easy to see that when these standards are met there is an extremely high chance that there will be no cable failure in the system. Although wires break, connectors are bent, and the unexplained things happen, if we build our cables according to a rigorous standard such as that specified by NASA we are giving our systems and our music the best chance to be heard and understood.

Visit the following websites to get a glimpse into what it takes for a cable termination to endure space travel, which is admittedly nothing compared to the same cable enduring a rock guitarist! http://nepp.nasa.gov/index.cfm/5575

SPEAKER CABLES

Use the appropriate wire to connect your speakers to your power amp. Speaker wire is not the same as a guitar cable. Use designated speaker wire. Also, choosing wire that is the wrong thickness for your situation can cause a problem with the efficiency of your amp and speakers. Speaker cable must contain two identical insulated wires and doesn't require a braided shield. The two wires in the speaker cable are typically composed of many strands of thin copper wire.

Sometimes, it is difficult to tell a speaker cable from an instrument cable because they both are commonly constructed with visually-similar wire and 1/4--inch tip-sleeve phone plugs on each end. Although speaker connections can be made with wire that looks like heavy-duty lamp wire, most professional speaker cable is round and black, like an instrument cable. These cables are not interchangeable. When a speaker cable is used as an instrument cable, the lack of a shield results in increased level of electrostatic noise and RF interference. When an instrument cable is used as a speaker cable, there is an impedance problem because the hot lead and shield are not the same impedance—like the pair of identical wires required by a speaker connection. The impedance variance will cause a decrease in the efficiency of the speaker-to-amplifier connection.

If there is a question about the intended application for a particular cable, open up the connector and inspect the wire. If there is an insulated wire and a

Chapter 2 .... CONNECTING

5 5

Wiring Differences

Notice that both jacks below look the same; however, one is designed to carry amplified signals to a speaker cabinet and the other is designed to carry line-level signals, such as those that come from a guitar or keyboard.

Speaker Cable

The speaker wire above contains two identical leads that connect to the tip and sleeve of the 1/4-inch phone plug.

The instrument cable below contains a single hot lead, which carries the instrument signal. The braided shield that surrounds the entire wire is twisted together and soldered to the tab connected to the connector's sleeve.

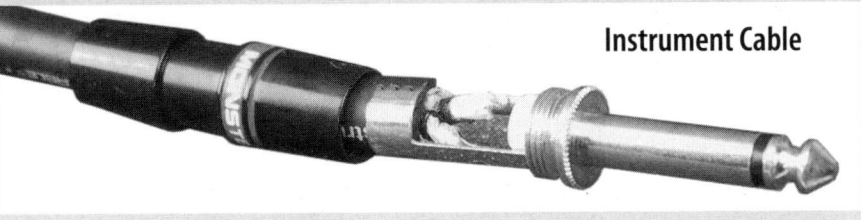

Instrument Cable

braided shield, it's an instrument cable. If there are two identical insulated wires, it's a speaker cable.

Wire Gauge

Speaker wire is categorized according to a few different wire gauge standards, the most common of which was established by Brown & Sharp, a division of Hexagon Meteorology, Inc. Brown & Sharp establish the American Wire Gauge (AWG) specification, which divides wire into AWG gauges from 0000 (pronounced "4 aught") gauge, with a diameter of .46 inches, to 36 gauge, with a diameter of 0.0050 inches. The actual gauge spec is based on a mathematical equation and it's not uncommon to see gauge tables from 00000000 (8/0) to 50 gauge—neither extreme applies to our duties as sound operators.

Wire exhibits impedance characteristics—different sizes of wire resist the flow of electrical current by different amounts. Larger wire exhibits less resistance

Microphones & Mixers... by BILL GIBSON

Speaker Wire Connections

Always use heavy-duty wire designed specifically for use with speakers. Wire is sized according the American Wire Gauge (AWG) standard. The smaller the wire number, the thicker the wire. Thicker wire has less resistance to signal. To have minimal degradation of signal in longer runs, we use thicker wire.

Be absolutely certain that the red post on the back of the power amp is connected to the red post on the back of the speaker and that black goes to black! If these are connected backwards on one of the speakers, the speakers are said to be out of phase.

When this happens, a sound wave that is sent simultaneously to both speakers moves one speaker cone out while it moves the other speaker cone in. Speakers connected out of phase work against each other instead of with each other. What you hear from them is inaccurate and unpredictable, especially in the lower frequencies.

(lower impedance) and smaller wire exhibits greater resistance (higher impedance). It's important to note that inappropriately small wire used for excessively long runs could add substantially to the combined speaker/cable impedance load. Using the appropriate cable length and gauge should result in a small impact on the overall impedance of the connection between devices. The actual impact of the cable on a system depends a variety of factors, including the cable length and thickness, the amount of power being provided, and the total impedance of the speaker cabinet. With all of these things considered, most speaker cable runs in a live sound setting require between 18- and 10-gauge wire. Some engineers prefer smaller than typical gauge cable (larger than typical diameter wire) in an effort to minimize the impact on overall impedance. In addition, thicker cable is often associated with better transfer of low frequencies. If you're building a large system and you're not exactly sure which wire is appropriate, consult an experienced and reputable system designer who has a track record of first-class installations similar to yours.

Chapter 2 .... CONNECTING

57

INSTRUMENT-LEVEL CABLES

Instrument cables are almost always designed for unbalanced connections. They use one hot lead containing the instrument's output signal surrounded by a braided shield, which helps absorb, reject, and diffuse electrostatic noises and RFI (radio frequency interference). Instrument cables typically utilize 1/4-inch TS phone connectors on each end.

Some speaker cables look similar to instrument-level cables; however, they use two completely different types of wire and are not interchangeable.

LINE-LEVEL CABLES

Line-level cable is designed to carry signals like those from a keyboard or guitar to a mixer or instrument amplifier. Line-level cables also connect outboard gear to the mixer or the mixer to the power amplifier or the powered monitors. Line-level cables can either be unbalanced, with a tip-sleeve connector, or balanced, using XLR or tip-ring-sleeve connectors.

MICROPHONE CABLES

Microphone cables typically utilize two-conductor shielded cable with XLR connectors at each end. In the professional studio, microphone cables are often used to connect line-level devices, such as mixers, effects, and power amplifiers.

Mic Cables

Standard microphone cables have an XLR connector on each end. They use cable that contains a braided shield surrounding two hot leads. This connection to the mixer is balanced.

In addition, standard mic cables are often used to connect between the balanced line-level mixer inputs and the line-level outputs from outboard processors, recorders, DAWs, instruments, and so on.

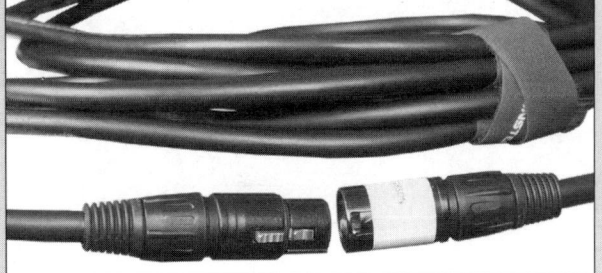

Devices that connect with microphone cables at line level use a balanced wiring scheme; devices that connect with regular guitar cables use an unbalanced wiring scheme.

DO CABLES REALLY SOUND DIFFERENT?

The difference between the sound of a poorly designed and a brilliantly designed cable can be noticeable. Exactly how extreme this difference is depends on the quality and type of the source material and the system through which it's being delivered. If a narrow-bandwidth signal composed of mid frequencies and few transients is compared on two vastly different cables, the audible differences might be minimal. However, when full-bandwidth audio rich in transient content, dimensionality, and depth is compared between a marginal and an excellent cable, there will typically be a noticeable difference in sound quality.

Listen for yourself. Most pro audio dealers are happy to show off their higher-priced product. When comparing equipment, listen to some fantastic recordings, with which you are very familiar, through the gear you're auditioning. It's best to use high-quality audio that receives industry praise for its excellence—after all, that's the standard you are trying to meet or beat. Right?!

Listen to Audio Example 2-3. The acoustic guitar is first miked and recorded through some common-quality cable. Then it's recorded through a microphone with some very high-quality Monster cable. Notice the difference in transient sounds, depth, and transparency.

· Audio Example 2-3
Mic on Acoustic Guitar Using Common Mic Cable, then Monster Studio Pro 1000 Cable

Audio Example 2-4 demonstrates the difference in vocal sound using common mic cable first, and then a high-quality mic cable from Monster. Notice the difference in transient sounds, depth, and transparency.

· Audio Example 2-4
Vocal Using Common Mic Cable, then Monster Studio Pro 1000 Cable

DIGITAL-INTERCONNECT CABLES

The cables you choose to make digital connections affect sound quality. You must use cable designed to carry digital data for transfers between digital devices. S/PDIF connections use RCA jacks and connectors. Even though a standard home audio RCA cable is capable of making the connection between a S/PDIF input and output, it is incapable of providing the stability and impedance required to facilitate an accurate data transfer. Similarly, an AES/EBU digital connection is made using XLR connectors. Even though a standard microphone cable is capable of completing the connection, a different type of wire is required to facilitate the most accurate data transfer.

Listen to Audio Examples 2-3 through 2-9. In each example, a different cable and format configuration is demonstrated. Listen specifically to all frequency ranges as well as transients. Also, consider the "feel" of the recording. Often, the factor that makes one setup sound better than another is difficult to explain, but it's easy to feel. The following examples are recorded using exactly the same program material and the identical transfer process.

The differences you hear on your setup depend greatly on the quality and accuracy of your monitoring system, as well as your insight and perception. Once you understand and experience subtle sonic differences, you'll realize the impact they hold for your musical expression. Constantly compare and analyze the details of your mixes. It will result in much more competitive quality. You'll realize more satisfaction, and you'll probably get more work.

I repeat: it's important to use cable designed for digital data transfer when connecting digital devices. Digital transfers require a specific and stable impedance of 75 ohms (S/PDIF) or 110 ohms (AES/EBU) to most faithfully transfer the digital data. A cable designed for analog applications transmits continuously varying voltage that mirrors the analog waveform it receives at the source. A cable designed for digital applications must efficiently and accurately transmit data, which is essentially a varying square wave indicating on or off ("1" or "0") binary bit status. These are two different tasks and they require different cables.

From the BlueJeansCable.com article "What Is Impedance, Anyway?"

> "Where analog audio or video signals consist of electrical waves, which rise or fall continuously through a range, digital signals are very different—they switch rapidly between two states representing bits, 1 and 0.

Microphones & Mixers.. by BILL GIBSON

This switching creates something close to what we call a square wave, a waveform that, instead of being sloped like a sine wave, has sharp, sudden transitions .... Although a digital signal can be said to have a "frequency" at the rate at which it switches, electrically, a square wave of a given frequency is equivalent to a sine wave at that frequency accompanied by an infinite series of harmonics—that is, multiples of the frequency. If all of these harmonics aren't faithfully carried through the cable ... then the "shoulders" of the digital square wave begin to round off. The more the wave becomes rounded, the higher the possibility of bit errors becomes. The device at the load end will, of course, reconstitute the digital information from this somewhat rounded wave, but as the rounding becomes worse and worse, eventually there comes a point where the errors are too severe to be corrected, and the signal can no longer be reconstituted. The best defense against the problem is, of course, a cable of the right impedance: for digital video or S/PDIF digital audio, this means a 75 ohm cable like Belden 1694A; for AES/EBU balanced digital audio, this means a 110-ohm cable like Belden 1800F."

Optical Digital Cables (Fiber Optics)

Fiber optic cables transmit bursts of light that represent the binary digital bits that represent audio sound waves. This type of cable is extremely well suited to the task and, conceptually, the cable should not make a difference in the data transfer. However, optical cable is fragile, doesn't reliably withstand bending, and, when subjected to harsh treatment, can't always be trusted to perform a flawless transfer. I have used a lot of optical connections in my studio, at other studios, and even in my own personal home entertainment system—I've had very few problems.

It's important to treat optical interconnect cables gently and to avoid bending them. Bending can cause cracks in the glass fiber material, which could degrade the light transfer and potentially damage, or interrupt the flow of, the data. Carefully route the cables to avoid kinks and bends; be certain there are no heavy devices setting on an optical cable. If you're careful to avoid bends and abuse to the optical cable, it should service as a very efficient and accurate means of making digital audio data transfers. In addition, optical connections are not susceptible to ground hums and RF interference.

In the audio world, there are two important optical protocols: TOSLINK and ADAT Lightpipe.

♦   TOSLINK: The term "TOSLINK" stands for Toshiba Link because Toshiba developed this two-channel stereo data protocol. TOSLINK data, carried via fiber optic cable, is essentially the same as S/PDIF data, which is carried via coaxial cable—inexpensive real-time converters from optical TOSLINK to coaxial S/PDIF are readily available. TOSLINK connections are commonly available on CD and DVD players, home entertainment systems and receivers, and most other home and professional stereo digital devices. In fact, TOSLINK connections are available on the headphone and microphone jacks on Apple's MacBook Pro. While TOSLINK can also carry multiplexed 5.1 DTS or Dolby Digital audio, it cannot carry six discreet digital audio channels.

♦   ADAT Lightpipe: Developed by Alesis for data connections between its ADAT digital recorders, ADAT Lightpipe (ADAT Optical Interface) carries up to eight tracks of 48-kHz, 24-bit audio. Alesis also developed a scheme to carry 96-kHz, 24-bit audio, but two ADAT Optical audio channels are required for each 96-kHz channel. ADAT Optical connections have been commonly adopted by the audio industry as a means to expand DAW

Cables for Digital Connections

Always use high-quality cables designed for data transfer when making connections between digital devices. Even though digital connections are sometimes made using XLR or RCA phone connectors, standard audio cables that also use these types of connectors present an incorrect impedance load for use in digital audio data transfers.

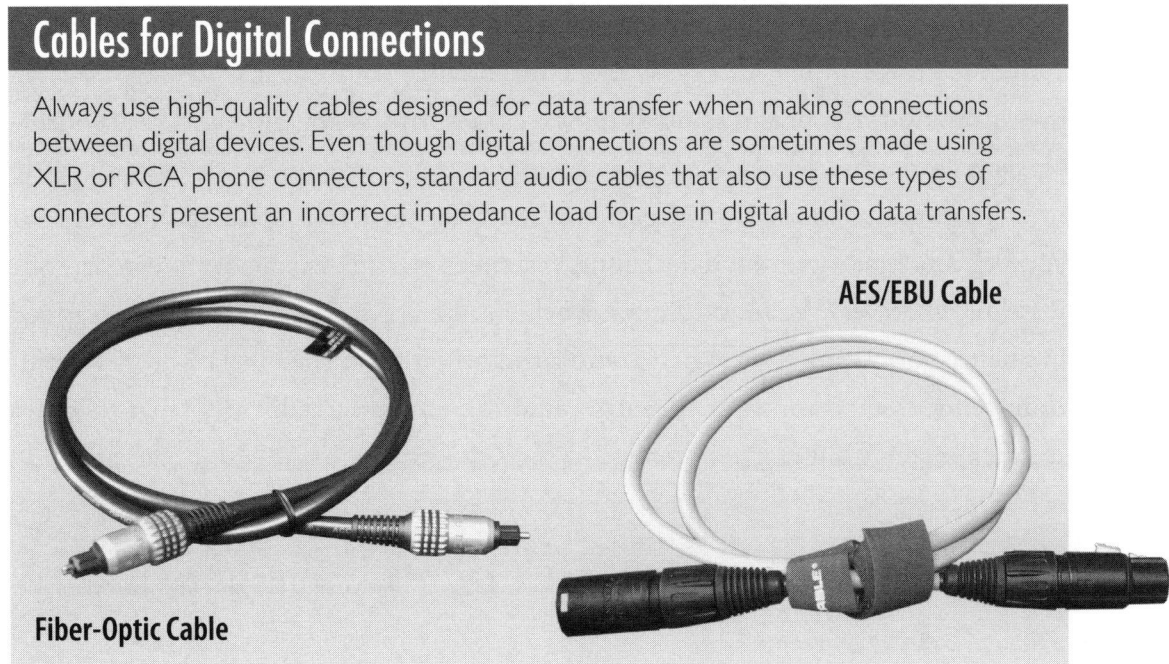

AES/EBU Cable

Fiber-Optic Cable

interfaces, eight channels at a time, and on digital mixers as a means to connect to various multitrack devices.

TOSLINK and ADAT Optical formats use the same connector and cable. These devices can be easily connected together, but to no avail—their respective data formats are incompatible.

Camp Time

There is a camp in the audio world that stands steadfastly to the theory that cables don't make a difference in digital transfers—it's usually the same camp that says a clone of a digital file can't possibly deviate from the original. Both of these viewpoints can be backed up theoretically and both have been disputed artistically.

In 1993, I brought highly respected and award-winning mastering engineer Bernie Grundman to Seattle to teach a seminar on mastering. Bernie services the highest-level clients in the recording industry and has won countless awards for his mastering work. With the digital age well on its way to ramping into modern recording culture (DAT was a standard and Pro Tools was first released in 1991), the hype about digital recording centered largely around the lack of noise and the value of identical digital clones. However, Bernie, who built his career on maintaining the highest standards in everything audio, wasn't willing to simply buy into the hype—he was hearing a difference between digital generations. He brought examples of cloned files and files that had been cloned multiple times to the seminar and played them for the audience. Tannoy had provided an amazing system for the seminar and everyone in attendance could hear the image collapsing throughout the progression of digital clones. It was amazing and very controversial at the time.

My point in sharing my experience with Bernie is to inspire you to listen for yourself and make your own decisions, whether you're assessing cable, clones, software, or whatever. We function in a world where art and expression are wrapped in science and technology. Artists tend to stand on what they feel or hear. Technicians tend to stand on what they can quantify mathematically and what they can see on a meter. That can get messy.

• Audio Example 2-5

AES/EBU to DAT Using Common Cable, and then SP1000 AES Silver Digital Monster Cable

Chapter 2 .... CONNECTING

63

· Audio Example 2-6

S/P DIF to DAT Using Common RCA Cables,
and then M1000 D Silver Digital Monster Cable

· Audio Example 2-7

Analog Out to DAT Using Common XLR Cables, and then Prolink Studio Pro 1000 XLR Monster Cables

· Audio Example 2-8

ADAT Lightpipe into Digital Performer Using Common Optical Cable, Bounced to Disk

· Audio Example 2-9

ADAT Lightpipe into Digital Performer Using Monster Cable's Interlink
Digital Light Speed 100 Optical Cable, Bounced to Disk

STUDIO REFERENCE MONITORS

Studio monitors are reference speakers, designed for specific applications. They provide an accurate audio image at a fairly specific distance and at a normal reference volume. They are different than home entertainment stereo speakers. If you use a typical home stereo speaker as a reference monitor you'll almost always get a sound that's unreliable. Stereo speakers are designed to sound good in a room; reference monitors are designed to provide an accurate image in a normal mix environment.

Selecting speakers is fundamental to producing excellent sounds that reliably transfer from your system to a friend's system, your car stereo, the radio, television, or any other performance medium. One of the most annoying and frustrating audio recording problems is a mix that sounds great on your system but sounds terrible everywhere else.

Part of the solution to this is experience. If you listen, analytically, to enough award-winning music on your system, you'll probably learn to match the basic sound. An even bigger part of the solution to this problem lies in the use of accurate and dependable near-field reference monitors. Industry standards are continually changing, and the market for near-field reference monitors has become very competitive. There are great new products available from all major speaker

64

Microphones & Mixers.. by BILL GIBSON

manufacturers, and many are very reasonably priced (typically between $300 and $1,000 per pair).

Speaker Cabinet Components

In order to effectively choose monitors and to assess their functionality it's important to understand a few things about how they work and what goes into them.

There are only a couple different types of components found in the majority of studio monitors. They're commonly referred to as *horns* and *cone speakers*, but more specifically, they are *direct radiator drivers* and *horn compression drivers*. Most monitors utilize some combination of these designs.

Drivers

The driver is the actual device that moves the air in a speaker. A speaker is a form of transducer—transducers simply convert one form of energy into another. The driver converts electrical energy from the power amplifier into acoustical airwaves.

Interestingly enough, the fundamental mechanical process that we use today to convert electrical impulses into airwaves was patented by Ernst W. Siemens in

The Cone Speaker

It is instructive to understand the cone speaker and its components. A moving-coil microphone and a cone speaker operate according to the same principle. At the core of both devices is a coil of copper wire suspended around a magnet. In a microphone, acoustic audio waveforms vibrate the diaphragm, connected to the copper coil, which causes a variance in the status of the magnet, which moves the speaker cone —this is the source of the audio signal.

In a speaker, the electrical signals from the power amplifier continuously vary the electromagnet at the core of the speaker, causing movement of the coil around the magnet and the attached speaker cone—this is the source of the variations in air pressure that we perceive as sound.

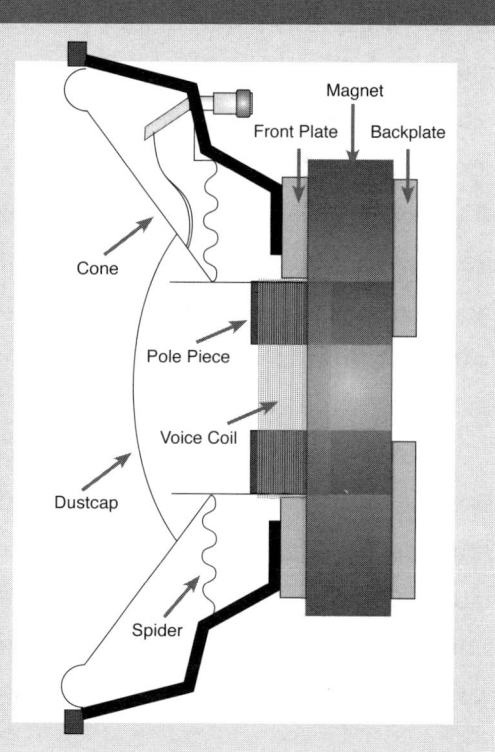

Chapter 2 .... CONNECTING

65

1874. This system utilizes a copper coil of wire suspended around an electromagnet. The magnet polarizes in response to the amplified audio signal, causing the suspended coil to move axially in response to the continually varying north and south magnetism. Axial movement involves a motion in two directions, such as in and out, up and down, or side to side. Though there have been refinements of the process and technical advances that have provided exceptional materials, this is still the way most speakers work.

The home theater and hi-fi industry offers some very nice new-technology speakers that sound incredible; however, most studio monitors still use a combination of horns and cone speakers.

Direct Radiating Cone Speakers

The cone speaker typically provides the low- and mid-frequency components of the audio signal, although the actual effective frequency range is dependent on the materials used and the diameter of the speaker. Commonly used direct radiating cone speakers vary in size from as small as a couple inches to as large as 18 inches—small speakers are more capable of reproducing high frequencies, and large speakers are more capable of producing low frequencies. Horn drivers are most efficient at handling high-frequency content of the audio signal.

Cone speakers need an enclosure to realize their low-frequency output potential. In space, without an enclosure, their response characteristic suffers from back-to-front cancellation. The energy created by the front of the speaker cone is, in reality, 180 degrees out of phase with the energy created by the back of the speaker—while the front of the speaker pushes the air in front of the cone, the back of the speaker pulls from the air behind the cone.

The principles of diffraction indicate that sound bends around obstacles, especially low-frequency sounds. The air movement from the back side of the speaker cone diffracts around the speaker and combines with the air movement from the front side of the cone in a destructive phase relationship. In fact, low-frequency waveforms are so much longer than the length of the path from behind the speaker cone to the front of it that phase cancellation approaches 180 degrees.

Horn Compression Drivers

Horn compression drivers utilize the same core functionality as the cone speaker. They both use a voice coil around a magnet to instigate the movement of air; how-

Microphones & Mixers .... by BILL GIBSON

ever, the compression driver is loaded down by air pressure created by a horn lens. A phase plug is used to help dissipate heat and lower distortion.

Compression drivers are more efficient than cone speakers and they are typically used to reproduce upper midrange and high frequencies. These drivers require less power to achieve the same SPL as cone drivers. Whereas a low-frequency cone driver might need to receive 1,000 watts for optimal performance, a high-frequency compression driver might only require a few hundred watts.

Impedance ratings for compression drivers are in the same range as the large cone drivers—they tend to fall in the range of 4 to 8 ohms.

Horns can be used both on cone speakers and on compression drivers. The opening of the driver influences the potential frequency response character—larger openings are more able to reproduce lower frequencies.

Ribbon Drivers

Some manufacturers use a ribbon driver for high and mid frequencies. These drivers are very similar in design to the ribbon microphone capsule, much the same way that the moving-coil mic capsule is like a cone driver. Ribbon drivers are capable of sounding extremely smooth and clean in the high-frequency band. When combined with an excellent low-frequency drive in a well-designed enclosure, these ribbon-based reference monitors can provide amazing audio quality that exhibits minimal distortion, which is easy on the ears throughout long mixing and tracking sessions.

The Importance of Enclosures

The capabilities of the drivers combine with the physical design of the cabinet to create a device with sonic characteristics. High frequencies rely very little on the enclosure dimensions and physical construction; however, low frequencies are radically colored, molded, and shaped during the design process. The cabinet size and construction often provide exaggerated pathways within the enclosure, which allow the low-frequency information to develop more fully. These pathways channel the low-frequency energy, focusing and enhancing it to the desired balance in relation to the mid and high frequencies.

The Cone Driver Outside the Enclosure

The enclosure plays a key role in realizing the deep tone that the cone driver is capable of producing. Air moves in response to in and out motion of the cone, creating the acoustic sound. As the cone moves outward, the air in the front of the speaker compresses—as the cone moves inward, the air in front of the speaker expands (compression and rarefaction). With the speaker in open air, there is one important

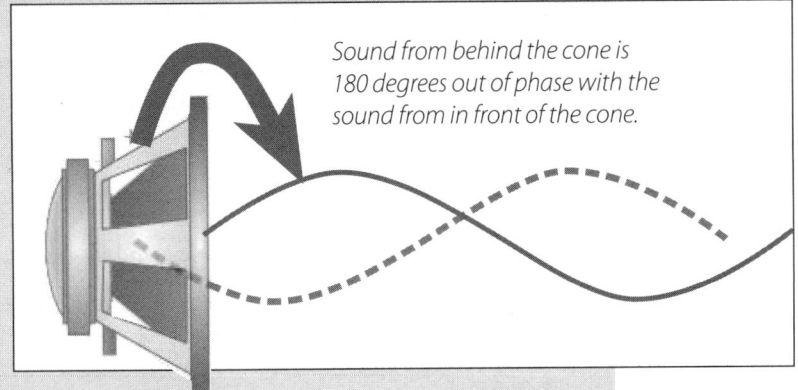

Sound from behind the cone is 180 degrees out of phase with the sound from in front of the cone.

problem—each time the cone pushes outward in front, it also pulls on the air behind the cone. This results in the simultaneous creation of two waveforms that are 180 degrees out of phase.

Because the low frequencies are very long, the distance from the back of the speaker to the front is insignificant; therefore, frequencies longer than about one foot tend to cancel, resulting in a thin and unreliable sound. Placing the speaker in an enclosure effectively removes the rear waveform information from the equation, allowing the low frequencies to reproduce accurately.

Sealed Enclosures

Sealed enclosures eliminate diffraction by closing off the path by which the rear-generated sound wave can combine with the front-generated sound wave. Absorption within the enclosure helps minimize destructive interactions between sound waves within the enclosure and the movement of the cone creating the sound being projected from the enclosure.

Infinite Baffle

The infinite baffle design utilizes a very large sealed enclosure in an attempt to take advantage of the resonance within the enclosure. These cabinets provide a very deep tone when properly designed, but they aren't typically known for providing a very flat or controlled frequency response characteristic.

Acoustic Suspension

Acoustic suspension enclosures utilize a relatively small enclosure, which is sealed to eliminate diffraction. The inside of the cabinet is lined with absorptive materials to minimize the destructive influence of mid-frequency reflections within the cabinet and help decrease resonance. This design is much more linear in its response in comparison to the infinite baffle design; however, it suffers a decrease in efficiency due to the small enclosure size.

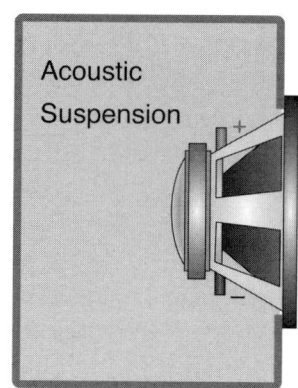

Bass Reflex

A bass reflex cabinet attempts to release the audio energy generated inside the enclosure in a way that augments its low-frequency response rather than decreasing it. In this design, an opening in the enclosure, called a *bass reflex port* or just a *reflex port*, lets the energy from inside combine with the energy being projected from the front of the speaker. If the enclosure dimensions are just right, and if the internal sound has been routed and released in the correct way, the bass reflex cabinet offers an excellent low-frequency response characteristic while providing a smooth midrange response.

Absorption is used within the enclosure to minimize destructive interactions from midrange reflections within the enclosure. The actual response impact created by the port is a result of the size of the port and the length on the pathway that the waveform must travel to reach its exit.

The port, a tube extending into the enclosure from the opening in the front of the cabinet, acts much like a flute. When air is blown across the hole, a tone is produced. A small flute (a piccolo) produces higher frequencies than a large flute (a bass flute). The same concept applies to the response of the reflex port in a speaker enclosure—the larger the port and the longer the tube, the lower the frequency produced.

The resonance of the port can make a reflex cabinet sound boomy at a specific low frequency. Ideally, the port is tuned so that it supports and reinforces the frequency-response deficiencies created by the enclosure's physical size.

The Passive Radiator

The passive radiator design utilizes a "dummy" (passive) speaker in addition to the active speaker. This passive speaker has no magnet or coil and it doesn't receive a powered audio signal. Its purpose is to support the motion of the active speaker by moving additional air in response to the activity from the air movement created by the active speaker. This design tends to result in a dip in frequency output at the enclosure's resonant frequency.

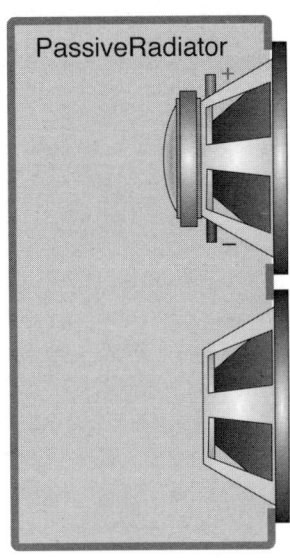

PassiveRadiator

In relation to the bass reflex cabinet design, the passive radiator can potentially provide an excellent low-frequency response in a smaller enclosure.

Transmission Line

The transmission line concept utilizes an extended-length port in an attempt to shift the phase of the sound waves generated by the rear of the speaker cone by at least 90 degrees. This is the type of system used by some manufacturers of small devices with unnaturally excellent bass response.

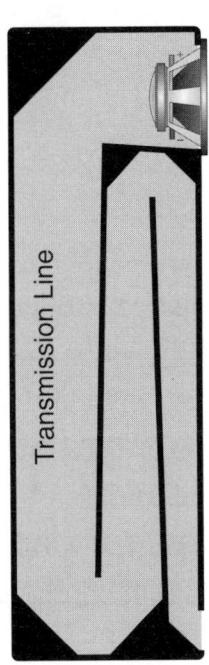

Transmission Line

A physical pathway is constructed within the enclosure—its purpose is to delay the escape of the sound waves inside the enclosure. Ideally, it will be delayed by an amount sufficient to allow it to combine in a constructive way with the sound wave emanating from the front of the speaker cone.

Component Alignment

Speaker enclosures that contain separate high-, mid-, and low-mid-frequency components must accommodate the size difference between the individual drivers. If all drivers are mounted on a flat surface—and if they are fed the same audio signal—there is a phasing discrepancy between the component outputs.

The physical size variation between tweeters, midrange, and low-frequency drivers results in a staggered point of origin for the audio that projects from each device. Because the sound doesn't originate from the same vertical plane, any

70

Microphones & Mixers.. by BILL GIBSON

Driver Alignment

Often, the drivers are simply mounted flush on the front of a simple, box-shaped enclosure—this is not always optimum. The physical alignment of the voice coils is sometimes adjusted by changing the shape of the enclosure. This is done so that there can be a more cohesive phase relationship through the overlap frequencies at the crossover point. Aside from physical positioning of the components, alignment adjustments can also be created electronically through delay circuits. Sometimes, speakers are shifted to compensate for inherent delays in the crossover circuitry.

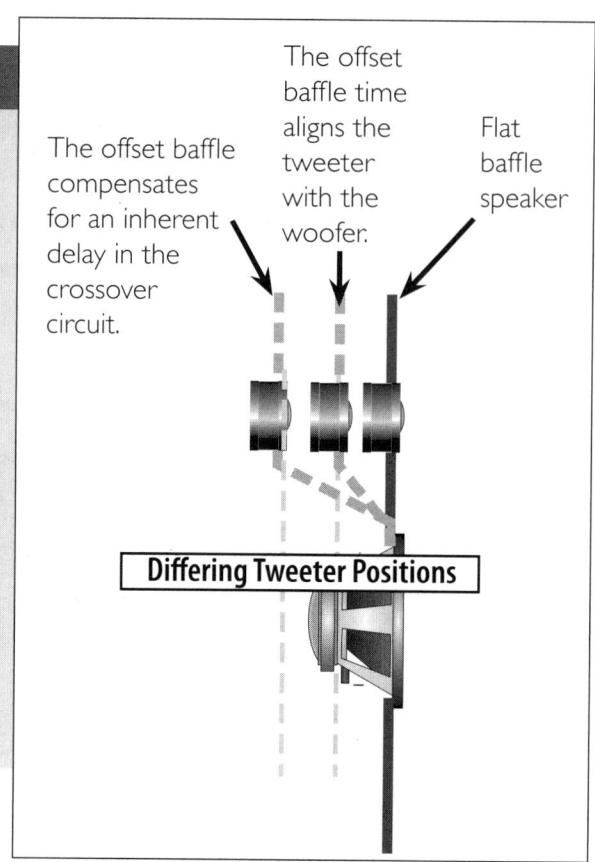

The offset baffle compensates for an inherent delay in the crossover circuit.

The offset baffle time aligns the tweeter with the woofer.

Flat baffle speaker

Differing Tweeter Positions

shared frequencies, such as those that reach into the crossover point, will be out of phase at the frequency that equals the length of the physical displacement.

In addition, timing incoherence between the frequency bands allocated to each driver blurs the audio image. Sounds that contain bandwidth that takes advantage of multiple drivers won't be reconstructed accurately if their content isn't reconstructed in a perfect phase relationship. The goal in any speaker design is to faithfully reproduce a broad bandwidth transient. A design has audio integrity if the transient audio signal is divided into frequency bands, through the crossover network, and then is reproduced by the components in the same phase relationship as the original source transient.

The process of positioning speakers so that the sound they produce is identical in phase to the source audio signal is often referred to as *time alignment, phase alignment, phase coherence,* or *transient alignment.* Shifting the audio output from each individual driver can be accomplished through physical design of the enclosure or through electronic delays.

Enclosures in which the tweeter sets back farther than the midrange driver are simply positioning the tweeter driver in line with the midrange driver. This design is effective and keeps the electronic circuitry simple.

Rather than creating an enclosure with the tweeter driver physically lined up with the midrange driver, some manufacturers simply mount the drivers on a flat surface, delaying the output of the tweeter so that it starts just as the signal from the midrange driver is on the same plane as its driver. Some manufacturers offset the physical position of a driver to compensate for a delay that is inherent in a particular crossover circuit.

Precise alignment is very important in the development of near-field reference monitors for use in the recording studio because the focal distance is very specific, typically within a meter or two. In these designs the sweet spot is along the same plane as the recording engineer's ear—standing up or changing position can radically alter the perceived speaker response.

Speaker Impedance

Speakers exhibit impedance characteristics, which must be matched to the amplifier outputs. Amplifiers are rated according to the load range of the source. Most amplifiers specify minimum operational impedance. The components of a speaker box are wired together in a way that produces the desired load (the impedance at the input enclosure's connector).

Impedance math is pretty simple. Each amplifier is rated at minimum impedance, which is typically 4 or 8 Ω—many modern amplifiers are rated down to 2 Ω. You simply need to be aware of the impedance load you are putting on the amplifier to optimize the power output and rating. Never connect speakers in such a way that the impedance load is less than the specified minimum. If the speaker offers too little resistance to the amplifier signal, it's a little like trying to push a little water through a big pipe—it can't sustain the push and eventually will overheat and fail. Proper impedance matching provides a balance, which results in the efficient transfer of power.

Calculating Speaker Impedance

There are two ways to wire two speakers together: in parallel and in series.

+   Multiple speakers wired in parallel present a lower impedance load to the amplifier output. To calculate the resulting impedance, use the equation

72

Microphones & Mixers... by BILL GIBSON

Parallel and Series Wiring

Two speakers, equal in load and wired in parallel, cut the impedance load in half. Two speakers, equal in load and wired in series, double the impedance load.

Where "S" equals speaker impedance, the equation used to calculate the resulting impedance from a parallel connection is Load=S1 × S2÷(S1+S2).

The equation used to calculate the impedance resulting from a series connection is Load=S1+S2+S3...

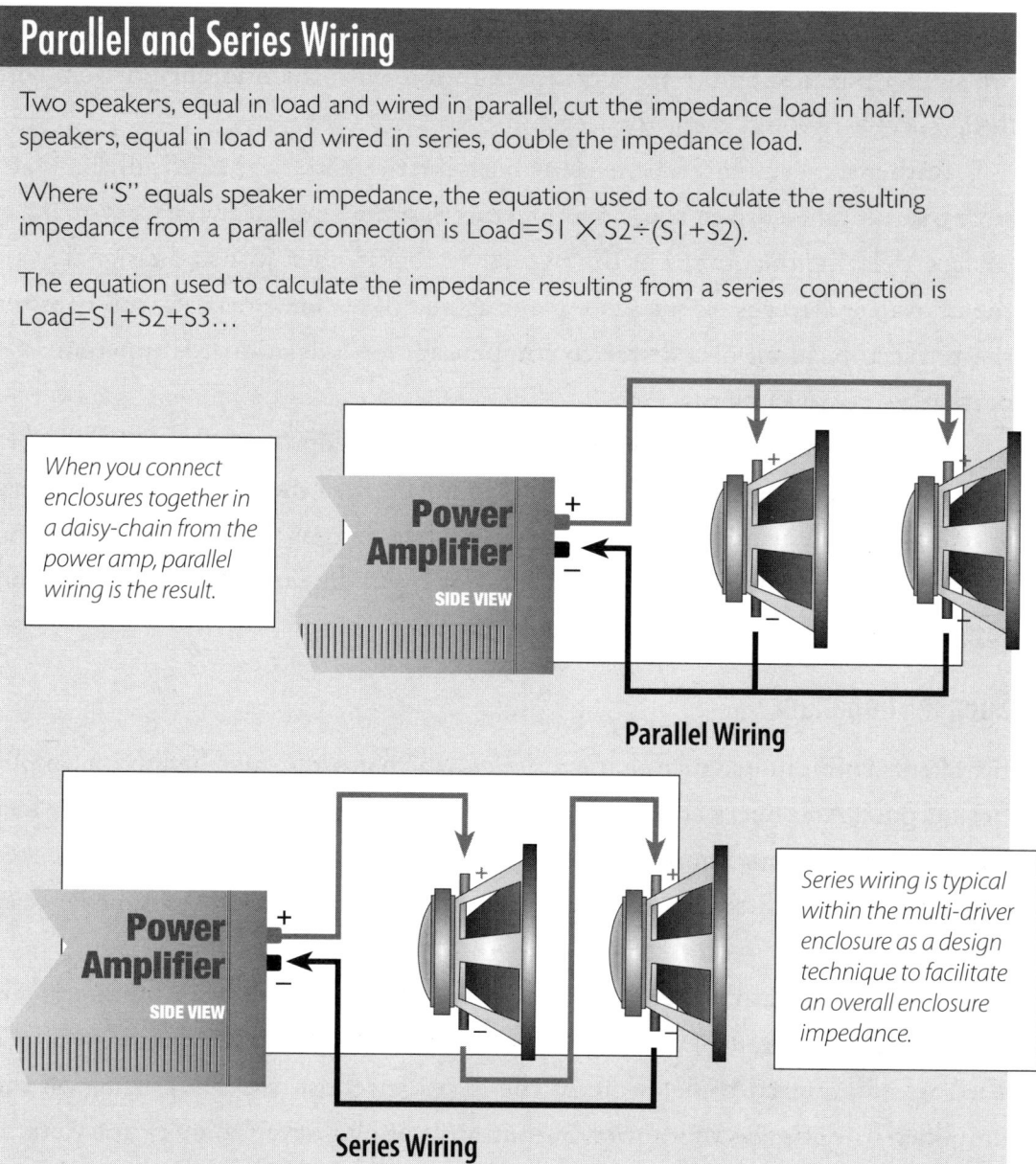

When you connect enclosures together in a daisy-chain from the power amp, parallel wiring is the result.

Parallel Wiring

Series wiring is typical within the multi-driver enclosure as a design technique to facilitate an overall enclosure impedance.

Series Wiring

$Z \, \Omega = (S1 \times S2) \div (S1 + S2)$ where S=Speaker Impedance. Two 8-Ω speakers wired together in parallel result in a 4-Ω load, $(8 \times 8) \div (8 + 8)$ or $64 \div 16$, which equals 4 Ω. Parallel connections are the most common between multiple speaker cabinets. This is the kind of connection you're making when you use a speaker cable to daisy-chain boxes, or when you stack dual banana connectors. In this scenario the positive post on each speaker input is connected to the positive post on the amplifier output. Also, both negative speaker inputs are connected to the amplifier's negative post.

Chapter 2 .... CONNECTING

73

Connecting Speaker Wire

To ensure correct speaker phasing, be absolutely certain that the red post on the back of the power amp is connected to the red post on the back of both speakers and that black goes to black! If these are connected backwards on one of the two monitors, the speakers are said to be out of phase. When this happens, a sound wave that is sent simultaneously to both speakers (panned center) moves one speaker cone out while it moves the other speaker cone in.

Speakers connected out of phase work against each other instead of with each other. What you hear from them is inaccurate and unpredictable, especially in the lower frequencies.

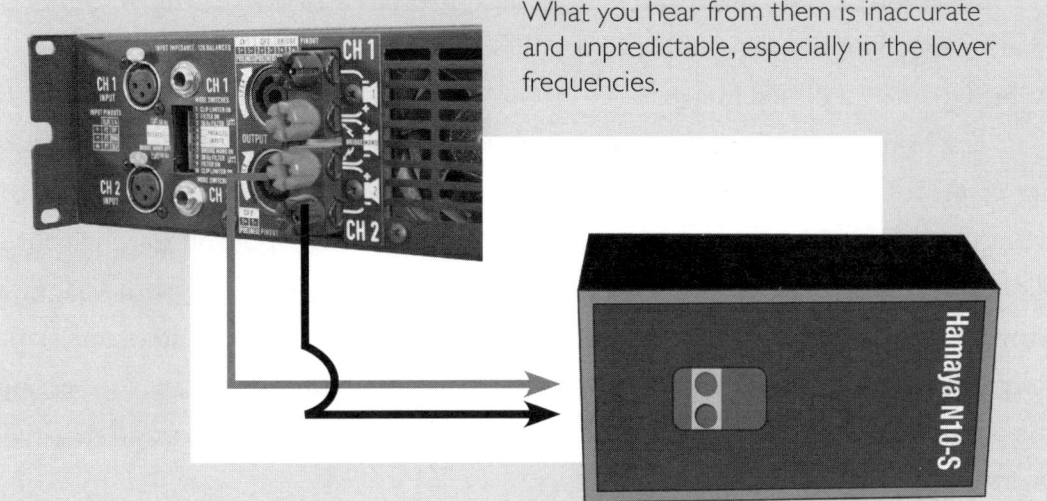

+   Series wiring is more common in internal speaker design using multiple drivers in a single enclosure, or whenever a large array is designed for certain applications. Multiple speakers wired in series present a higher impedance load to the amplifier output. To calculate the resulting impedance of multiple speakers wired in series, the load presented to the amplifier output is calculated as Z=S1 + S2 +S3 +.... Two 4-Ω speakers wired in series present an 8-Ω load to the amplifier. Three 2-Ω speakers wired in series present a 6-Ω load to the amplifier: Z=2+2+2. Series wiring completes a circuit from the positive post-amplifier output, through two speakers and back to the negative post on the amplifier. The positive post on the amplifier connects to the positive side of speaker A; the negative side of speaker A connects to the positive side of speaker B; then, the negative side of speaker B connects to the negative post on the amplifier.

BRIDGING AN AMPLIFIER

There are some instances, especially in system design where the audible frequency spectrum is divided into bands for delivery to optimized drivers and/or enclosures, where a stereo amplifier is transformed into a more powerful mono amp through bridging. Bridging combines the left-right output from the power amplifier into a mono output with increased power.

Be sure that the power amplifier is designed to be bridged. Any amplifier is easily bridged by simple connection modification, but if the amp isn't designed to be bridged it could be damaged.

The bridged connection is very simple to accomplish. From the power amplifier connect the positive post from output 1 to the positive post on the enclosure, then connect the positive post from output 2 to the negative post on the enclosure.

Some amplifiers provide a switch to select bridge mode with a diagram showing proper speaker connection. Amplifiers using banana connections typically position the outputs so that bridging is easily accomplished by connecting the plug across the left-right outputs instead of between the positive and negative posts on just one of the outputs.

Bridging the Power Amp Outputs

Be sure your power amplifier is designed to be bridged before you attempt this procedure. If your amp isn't specifically designed for bridging, it will be damaged.

Bridging the stereo outputs results in a mono output with increased power. The power amp below offers two sets of powered outputs: dual banana and Speakon. Using the banana connection in normal stereo mode, the channel one and two red and black terminals connect to the corresponding speaker enclosure terminals. Notice that in mono the single banana connector is flipped horizontally to connect to the positive terminals of channels one and two.

To bridge to mono, simply connect a banana plug to the red posts or select BRIDGE mode and connect to the Speakon MONO BRIDGE output.

Chapter 2 .... CONNECTING

75

CROSSOVER CONFIGURATIONS

Speaker cabinets typically contain components that reproduce specific frequency bands, ranging from the lowest frequencies to the highest. In addition, many systems are designed to include multiple speaker cabinets that each cover a specified frequency band from lowest to highest.

The full-range audio signal is split into frequency bands by an electronic circuit called a *crossover*. Typical crossovers divide the audio into two, three, or four bands. Each band is sent to a component (or components) in a cabinet or, in large systems, to separate cabinets optimized for a certain band. A separate power amplifier is typically used for each band. Therefore, each frequency band can be powered according to its requirements—low frequencies require substantially more power to accurately reproduce than high frequencies.

Using a crossover typically produces the best sound quality, but it must be adjusted correctly for the system components. It's important that the sound operator understands the concepts and considerations involved in the utilization of crossovers in audio system design.

Passive Crossover

There are two types of crossovers: passive and active. A passive crossover is typically built into the cabinet

Passive Crossover

The passive crossover receives a powered, full-bandwidth signal from the power amplifier. This signal is split into high- and low-frequency content through filters. Crossover frequencies and the severity of the filter slopes are carefully selected depending on the speaker components and intended application.

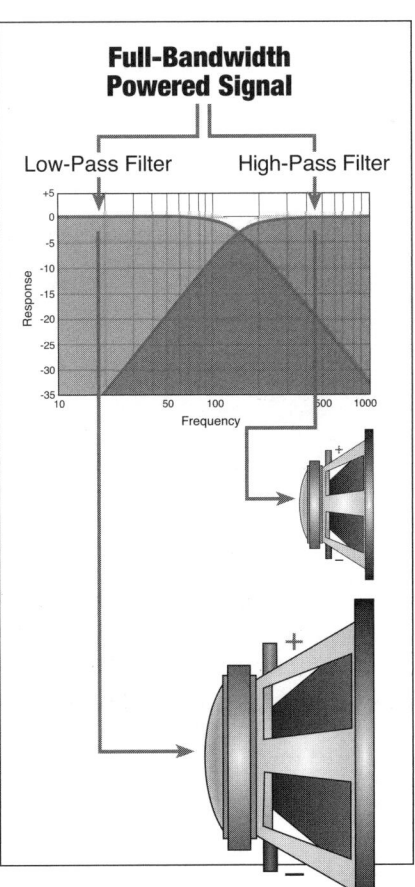

76

Microphones & Mixers... by BILL GIBSON

and receives a full-range signal from the power amplifier. High-pass, low-pass, and bandpass filters divide the frequency spectrum for delivery to the target components.

Passive crossovers don't require electrical power to operate—they are merely circuits that eliminate (filter out) specified frequency bands. These filters ramp up and down at different rates specified in decibels per octave—most commonly at a rate of 6 dB per octave (first order), 12 dB per octave (second order), 18 dB per octave (third order), and 24 dB per octave (fourth order).

Passive crossovers are typically less expensive than active crossovers and they don't require electrical power to operate. The fact that they receive their signal from a power amplifier is important. In order to withstand the sheer energy of a modern, high-wattage amplifier, the passive crossover circuitry must be substantial. Even if it is capable of withstanding a constant barrage of power over long periods of time—like from the beginning of a session to the end—the passive crossover's electronic components will usually heat up. As electronic circuits heat up, their response characteristics typically change, causing them to produce an inaccurate and uncharacteristic sound.

Active Crossover

Whereas a passive crossover, typically located inside the actual speaker enclosure, receives a signal from the output of a power amplifier, an

Active Crossover

Whereas the passive crossover receives a powered signal from the output of the power amplifier, the active crossover receives a line-level signal from the output of the mixer or signal processor. The line-level signal is divided into multiple frequency bands and then sent to the line input of separate power amplifiers.

Because the active crossover receives line-level signals rather than powered signals, there is less demand on its circuitry; therefore, the active crossover output is much more stable over the course of extended use than the passive crossover output.

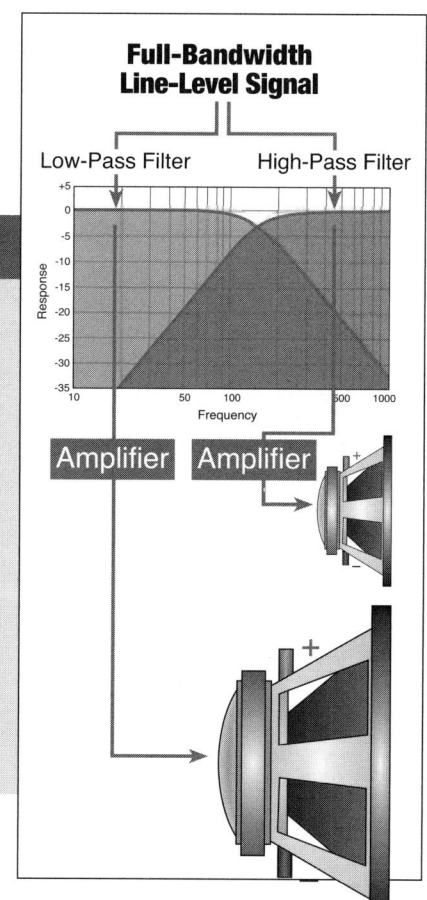

Chapter 2 .... CONNECTING

77

active crossover receives a line-level signal, from either the mixer or in-line processor. Active crossovers require electrical current to operate. The filters included in an active crossover are typically adjustable and accurate. These valuable tools let the system designer adjust the size and strength of each band.

The point where one band transitions to the next is called the *crossover point*. Systems comprised of custom-built or specifically selected components often require carefully chosen crossover points in order to realize the full system potential. Active crossovers provide ample control over the band selection and distribution.

Because the active crossover splits the frequency spectrum into multiple bands, keep in mind that each band will then require one power amplifier, or at least one channel of a stereo amplifier, to operate its target cabinet or component.

Full-Range Cabinets

Full-range cabinets use passive crossovers to distribute the incoming, powered signal to the individual components. Some speaker components act as full-range devices, but they are rarely used in high-quality professional sound systems.

Many cabinets accept a full-range powered signal and pass it through a passive crossover, while at the same time offering access to the individual components in the cabinet directly, bypassing the passive crossover. This affords the opportunity to divide the frequency spectrum through an outboard active crossover, then route the specific bands through separate power amplifiers, which are then connected to the individual components.

Two-Way Cabinets

Two-way cabinets, traditionally referred to as *bi-amplified*, contain components that are optimized for two specific frequency bands: highs and lows. The components determine the exact crossover frequencies. A two-way system is better than a full-range system, but it isn't the optimum in most circumstances. The high-frequency drivers require less amplification than the low-frequency drivers—a two-way system can adequately separate the highs, but the remaining band is so broad that it becomes inefficient as it tries to simultaneously push the mids and lows through the same component.

Two-way systems work fairly well in a near-field studio monitor application where the speakers are within about one meter of the listener's ears; large far-field systems often use three- or four-way configurations.

Three-Way Cabinets

Three-way systems are capable of providing excellent sound. High-quality components combined with excellent power amplifiers should faithfully reproduce a good audio source. Three-way cabinets provide access to components that are designed to reproduce highs, mids, and lows. Sometimes the low-frequency components are unduly stressed as they attempt to reproduce the low-mid, bass, and sub-bass frequencies at the same time.

Depending on the style of music and the desired sound pressure level, a three-way system could provide perfectly acceptable or woefully marginal results—it's really all about the sub-bass frequencies below about 100 Hz. These frequencies require the most power to accurately reproduce and they are the most adversely affected by the three-way design.

Four-Way Systems

Four-way systems typically add one or two subwoofers to a three-way system. The addition of components capable of receiving the sub-bass frequencies along with sufficiently powerful amplifiers provides a noticeable increase in sub-bass frequency content. At the same time, removing the sub-bass frequencies from the components that handle the low-mid and low frequencies radically increases their efficiency.

Most large high-budget systems use this configuration. Some very large systems split the frequency spectrum into more than four bands. Four bands with intelligently chosen crossover frequencies and ample power can adequately represent a well-mixed, full-range audio source.

POWERED AND NON-POWERED MONITORS
Non-Powered Monitors

A non-powered reference monitor requires an external power amplifier. The speaker enclosure is designed to work together with the speakers to provide an

Chapter 2 .... CONNECTING

79

accurate sound from a specified distance. Inexpensive models typically contain a passive crossover. This is the least efficient system.

Some non-powered monitors provide individual access to the high-, mid-, and low-frequency components. With this system, you're provided the option of utilizing an external electronic crossover to divide the full-range signal at line level. Once the frequency ranges are established, they're connected to an external amplifier, which is then connected to the appropriate speaker component. This is a very efficient system.

There is one critical advantage to non-powered studio reference monitors: the entire system can be customized to fit a specific need and environment. You get to select the monitor you love, combine it with the amplifier(s) you love, and

One-, Two-, Three-, and Four-Way Cabinets

Two-, three-, and four-way cabinet designs typically include a crossover for distributing the appropriate powered frequency bands to each speaker component. In addition, there are usually separate input jacks that provide access to each speaker if the sound operator chooses to use an active crossover to separate the line-level audio signal into separate bands.

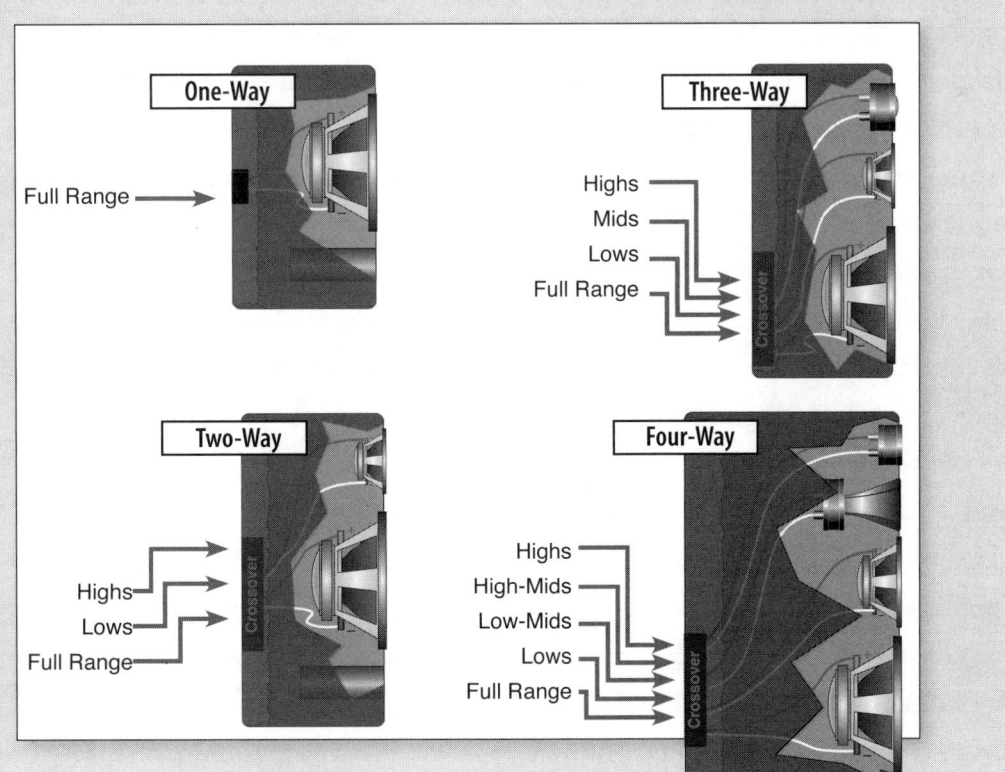

use the cable and connectors you love. This is a wonderful scenario, especially if you like to spend a lot of money.

There are a few disadvantages to using non-powered monitors:

- Cost: Especially if your system utilizes an electronic crossover with multiple power amplifiers, the combined cost is substantial.

- Efficiency: At home, where budget is a consideration, most people use one stereo power amplifier to send the left and right powered signal to the passive crossover, which is built into the speaker. This is the least efficient system, especially if you work for hours at a time on your music. The powered signal tends to heat up the crossover components, which, in the course of time, changes their operational characteristic. In other words, they sound different at the end of the session than they do at the beginning.

- Ambient noise: Many high-quality power amplifiers require a cooling fan. Most commercial studios keep noisemakers, like amplifiers, CPUs, and tape machines, in a separate room so the actual monitor environment is as free as possible from extraneous ambient noise.

Professional Amplifiers

Connecting the mixer to your power amp is an important step. Use high-quality cables to connect the outputs of the mixer to the inputs of the power amp. Use balanced line-level connections whenever possible. You'll get a better and more accurate signal that's free from radio interference and other noise. Many wires are specially designed for minimal signal loss. This means a better signal-to-noise ratio. High-quality wires and connectors also last longer and create fewer problems.

Using a quality power amp is very important. Distortion is a primary cause of ear fatigue, and an amplifier that produces less distortion over longer periods of time causes less fatigue and damage to your ears.

If you have a professional power amp with a rating of at least 100 watts RMS, and if you use a good quality reference monitor designed for studio use, you'll be able to work on your music longer with less ear fatigue. When I use the term "professional" in regard to amplifiers, I mean an industrial-strength unit, designed for constant use in a professional setting. Compared to a consumer device, amplifiers designed for professional use generally have better specifications, therefore helping to reduce ear fatigue. They use high-quality components, therefore lasting

Chapter 2 .... CONNECTING

81

longer and working harder for longer periods of time. Reputable manufacturers offer the best service and support. Fast, quality service is invaluable when you're making money with your equipment.

Using a small system designed for home entertainment is unacceptable for monitoring during any recording project. It's always a great idea to check your mix on a small home stereo system, to verify that it sounds good in that environment; however, if you want your music to sound good in a variety of listening environments, you need a good set of studio reference monitors and a trustworthy amplifier.

Self-Powered Monitors

A self-powered monitor has the amplifier built into the speaker cabinet. Typically, the amplifier is either contained within the speaker enclosure or it simply mounts to the back of it. This is very convenient and efficient.

Reference Monitors

Reference monitors have evolved over the years into sophisticated, microprocessor-based tools. For years, Yamaha NS10 near-field reference monitors have been one of the most popular and trusted tools in the recording engineer's toolkit. When they came out, in 1978, they were relatively inexpensive and easy to get. Engineers liked them because they were pretty accurate, but also because they tended to force the mix engineer to build a clean mix that sounded good on a lot of different systems and when broadcast over the airwaves. They're still a commonly used tool that remains on the meter bridge of many professional studios.

Reference monitor technology has come a long way. The JBL LSR4328P has a built-in processor and uses a calibrated microphone to listen to the speakers in the room, making adjustments to the frequency balance and acoustic response to provide the flattest and most trustworthy reference possible in your own room! The monitors connect together with CAT-5 cable so that every component, including the optional subwoofer, works together to provide a trustworthy image at the mix position. In addition, these monitors include a software interface that lets the user adjust balance and EQ settings manually.

Microphones & Mixers.. by BILL GIBSON

A number of years ago, self-powered monitors were very expensive and really not justifiable by most home recording enthusiasts. However, they were very reliable, efficient, and accurate. This is an excellent case in which manufacturing, market demand, and technology came together to make self-powered monitors available and affordable. Now, virtually all manufacturers make affordable self-powered studio monitors designed specifically for near-field applications.

There are several advantages offered by the self-powered monitor over non-powered models:

* They are very cost-efficient. Since all the components are designed to work together, the production process is streamlined, keeping manufacturing costs down and quality up.

* They are very easy to connect. Simply supply AC power and connect a cable from the mixer line-level output to the speaker's line-level input.

* They are very efficient. The amplifier is designed to match the speakers and the box.

* Since the amplifiers are efficient and small, they don't require cooling fans—they don't add to the ambient noise level. If you aren't cranking out the hits, they're silent.

* For two- and three-way monitors, the frequency ranges are divided (crossed over) prior to the amplifier stage. Once the crossover divides the full-range signal into two, three, or more ranges, those signals are sent to separate amplifiers, which in turn power the individual components. Whereas the passive crossover, in a non-powered monitor, changes characteristic when used continuously, the powered system provides a monitor that's consistent and accurate for long periods at a time.

There are few disadvantages to self-powered monitors. They're affordable, efficient, and they sound good. However, if you have an unlimited budget and lots of time to spare, there are a lot of wonderful components that you can combine for a truly amazing listening experience.

Power Ratings

Power ratings are expressed in watts. It is very important to know something about two types of watts in order to make any assessment about amplifiers.

PMPO stands for Peak Maximum Power Output, usually referred to as simply peak power. Peak power lasts only for short durations during momentary blasts of power. Manufacturers sometimes refer to the peak power rating to impress the prospective customer. Peak power ratings are fairly meaningless in our assessment of power amplifiers.

RMS (root mean square) power represents the average sustained power output. RMS is the important power specification for our consideration. The peak power rating is typically at least twice the RMS rating.

The Value of a Subwoofer

It's easy to get confused over the value of a subwoofer. Creating a big fat boomy bottom in the mix is not the purpose of the sub. However, it can definitely help us build an *accurate* big fat boomy bottom in the mix.

Subwoofers typically reproduce the frequencies below about 100 Hz. The exact crossover point is dependent on the response of the room and the components in your monitor system. These low frequencies require a lot from a small monitor and frequently detract from its ability to respond to the mids and highs.

Here are a few of the primary benefits of the subwoofer:

- Redirecting the low frequencies from the main monitors to the sub releases the burden on the main speakers so they can more efficiently reproduce the rest of the frequency band. The sonic difference is often stunning.

- Subs accurately provide low frequency content in the mix. If your monitor system is incapable of reproducing low frequencies, it's likely that, at some time, you'll build a mix that sounds great on your system but is extremely bassy and boomy on a full-range, high-quality system that is capable of reproducing the lows—never fun.

- Separate control over specific low-frequency interactions in the control room is easier to achieve when routed through the sub. Low frequencies are the most difficult band to control acoustically; redirecting them to the subwoofer offers more exact control in virtually any room. If everything is working correctly, it shouldn't be obviously noticeable that there is a subwoofer in the system until the mix engineer consciously—or unconsciously, for that matter—adds those frequencies.

Microphones & Mixers .... by BILL GIBSON

Power ratings are a logarithmic function. Since they follow a logarithmic scale, mental comparisons between power ratings are a little confusing at first. In order to make a 3-dB boost, twice the power is required. Since 1 dB is essentially the smallest audible difference in volume, 3 dB just doesn't add up to much, especially considering the power requirements to achieve it. To increase the level by 6 dB, a four-fold power increase is necessary.

Near-Field Reference Monitors

A near-field reference monitor should be used with your head at one point of an equilateral triangle (approximately three feet, or one meter, on each side) and the speakers at the other two points. The speakers should be facing directly at your ears and are ideally about 10 degrees above the horizontal plane that's even with your ears. With this kind of a system, the room that you're monitoring in has a minimal effect on how you hear the mix. These monitors should sound pretty much the same in your studio at home as they do in any studio in the world.

If the room is minimally affecting what you hear, then the mix that you create will be more accurate and will sound good on more systems. Changing to a near-field reference monitor gives you immediate gratification through more reliable mixes, plus it lets you start gaining experience based on a predictable and accurate listening environment.

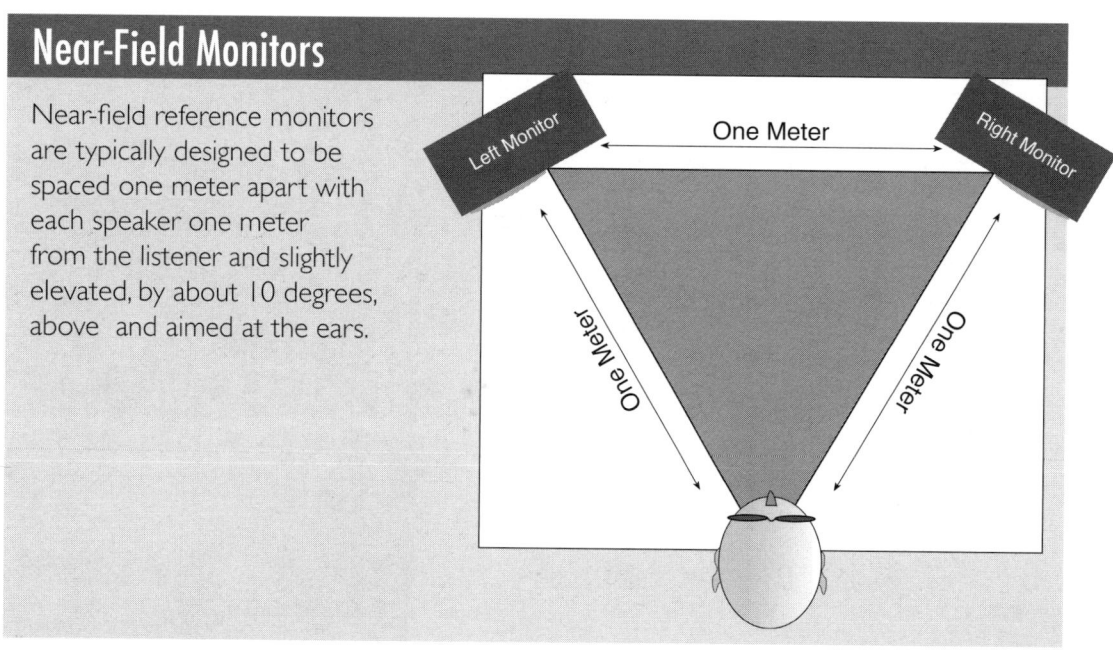

Near-Field Monitors

Near-field reference monitors are typically designed to be spaced one meter apart with each speaker one meter from the listener and slightly elevated, by about 10 degrees, above and aimed at the ears.

Chapter 2 .... CONNECTING

Not just any small speaker works as a near-field reference monitor. In fact, speakers that aren't designed specifically for this application produce poor results and unreliable mixes when positioned as near-field reference monitors.

Far-Field Monitors

Far-field monitors are designed to be farther away from the engineer, and their sound is greatly affected by the acoustics of the room they're in. Larger rooms have more air to move, so they require larger monitors to move that air. These monitors can be very expensive.

In order to get great results from far-field monitors, they must be used in a studio that has been acoustically designed for a smooth and even balance of frequencies within the room. Since this can involve actual construction and often plenty of expense, and since near-field reference monitors can produce excellent results, the obvious choice for most home setups is a pair of near-field reference monitors.

Far-Field Monitors

Most commercial studios, like the one pictured here, contain multiple monitor systems. It is important for the audio engineer to be able to listen to the sounds he or she is creating on different sizes of systems. Notice the far-field speaker as well as the near-field system. In addition, most studios keep a set of Auratone 5Cs around to be able to hear their mixes on a system similar to a radio or television broadcast. These systems are always listened to separately—they are never combined.

Dark Horse Studios • Nashville, TN

86

Microphones & Mixers.. by BILL GIBSON

PLUGGING IN

The output of your mixer might have multiple outputs for connection to different amplifier inputs. The Main output is the correct output from the mixer to be plugged into the power amplifier. This might also be labeled Mains, Mix Out, Out to Amp, or Stereo Out. If your mixer has XLR outputs available, and if your power amp has XLR inputs, patch these points together as your first choice. This output typically provides the cleanest and most noise-free signal. Many mixers and amplifiers use 1/4-inch connectors that are capable of producing or receiving either unbalanced or balanced signals. Use balanced connections wherever possible. If you need to use unbalanced connections, remember to use the shortest cables possible to avoid radio interference and extraneous noises.

Electrical Power

Spikes and surges are fluctuations in your electrical current that rise well above the 120-volt current that runs most of your equipment. Surges generally last longer than spikes, but both usually occur so quickly that you don't even notice them. Because power surges and spikes can seriously damage delicate electronic circuits, protection is necessary for any microprocessor-controlled equipment (computers, synthesizers, mixers, processors, sequencers, printers, and so on). Oh yeah—I guess that's virtually everything we use these days!

Inexpensive power strip surge suppressors are not the ideal solution to protection against spikes and surges, although for a small system they are better than no protection at all. For reliable electronic system protection, dedicated power conditioners guard against spikes, and some conditioners even maintain a constant voltage level. Furman offers several basic affordable options that are functional and efficient. They also offer more expensive systems that provide RMS voltage regulation, surge protection, noise filtration, voltage protection, voltmeter/ammeters, isolated outlet banks, USB chargers, and diagnostic lights. Especially when using a large complex system, it is worthwhile to invest in high-quality power supply and distribution systems.

Powering Up

+ Turn on the mixer and outboard gear (such as delays, reverberation devices, and compressors) before the power amps.

Chapter 2 .... CONNECTING

87

+ Always turn power amps on last to protect speakers from pops and blasts as the rest of the electronic gear comes on.

Powering Down

+ Turn power amps off first to protect speakers, and then turn the mixer and outboard gear off.

GROUND HUM

Aside from causing physical pain, grounding problems can induce an irritating hum into your audio signal. If you have ever had this kind of noise show up mysteriously and at the worst times in your sound system, you know what true frustration is.

.... Audio Example 2-10

60-Cycle Hum

Sixty-cycle hum is the result of a grounding problem in which the 60-cycle electrical current from the wall outlet induces a 60-cycle-per-second tone into the audio signal.

To make matters worse, this 60-cycle tone isn't just a pure and simple 60-Hz sine wave. A sine wave is the simplest waveform and, in fact, is the only waveform that has a completely smooth crest and trough as it completes its cycle. We could easily eliminate a 60-cycle sine wave with a filter. Sixty-cycle hum has a distinct and distracting waveform, which also includes various harmonics that extend into the upper frequencies.

It's very important to have your setup properly grounded in order to eliminate 60-cycle hum and for your own physical safety while operating your system.

Grounding

Grounding is a very important consideration in any setup! The third pin, called the ground pin, on an AC power cable exists for your safety. It is a key component in a scheme designed to keep you from being electrocuted. If there's an electrical short or a problem in a circuit, the electricity may search out a path other than the one intended. Electricity is always attracted to something connected to the ground we walk on (the earth). The ground pin gives an electrical problem such as this somewhere to go. It provides a low-resistance path for the dissipation of fault current.

If a piece of equipment has a ground connection that is intermittent or disconnected and if you happen to touch that equipment, you might become just the path to the ground that the electricity is looking for. This could be painful, at the very least, or even fatal. Properly grounding a piece of equipment gives potentially damaging electrical problems a path to ground other than you.

Ground Loops

Grounding, while providing essential safety, offers potential signal noise through the formation of ground loops. Any time we connect multiple electrical components together (outboard equipment, mixers, amplifiers, and so on), we run the risk of creating a ground loop. This is actually a loop between multiple ground circuits, which essentially acts as an antenna, receiving and absorbing interference and noise that becomes infused in our intended audio signal.

If a system is designed properly, there shouldn't be a grounding problem and, in fact, the grounding scheme will help eliminate noise from entering the system; however, a sound operator on the road doesn't always have the luxury of integrating with a properly designed audio or AC system.

In the audio world, ground loops cause hums and buzzes. In the video world, they cause interference bars in the picture. Functionally, ground loops often cause erratic operation or even damage to audio and video equipment.

Solutions to Grounding Problems

Let's look at some practical solutions to the persistent hums and buzzes that vex so many setups. Once your sound system is on the right path electrically, your frustration level should drop significantly.

Seasoned designer/installers typically have the luxury of designing all aspects of a system and implementing a meticulously wired and impeccably integrated electrical, audio, and media system. In these installations, there is time to sleuth out buzzes and hums, and then there is time to implement a plan to eliminate them. That's not the quite the same thing as getting a band all set and ready to go for a session and finding yourself faced with an unbelievably loud hum, usually because of one stubborn piece of the band's gear. The recording engineer needs

to have some solutions already in mind and be able to track down and eliminate errant noises quickly.

Noise Doesn't Necessarily Indicate a Grounding Problem

A noise in the system might be coming from a bad cable. One of the possible scenarios for cable failure is that the shield becomes disconnected, which results in a noisy hum along with the intended signal. If you hear a noise in the system, first look at the channel, group, and main meters. Frequently, the offending noise will show itself clearly on one of the meters. If you see a channel meter that looks like it is receiving a constant input level, turn that channel off to see if the problem disappears. If the noise goes away as the channel is muted, replace the cable between the mixer or snake channel and the offending device. Or, if the noisy channel is connected to a direct box, push the ground button on the DI. If the noise disappears as a result of either of these procedures, celebrate and enjoy the session!

Connect All Equipment to the Same Outlet

This is a simple and safe solution to this common problem but it is only viable with small or mid-sized system that doesn't require multiple 15- or 20-amp AC circuits. Because the ground loop is caused by the differences in electrical potential (accumulated voltage) between outlets in your building, plugging into one outlet with all of your equipment drastically reduces the likelihood of grounding problems.

Reminder: Always use high-quality AC distribution. Plastic electrical power strips are known to present a fire threat, so use metal power boxes capable of handling your equipment. It's a good idea to hire an electrician or a system designer/installer to design and build an appropriate electrical distribution system.

Hum Eliminators

Ebtech makes both line-level and AC hum eliminators. These devices isolate the ground without disconnecting it. In many cases, they will rid the system of an annoying hum and the show can go on. Some designers are not fans of these devices and, for a permanent installation, hum eliminators are not the best solution. The best answer is to find the source of the noise and get rid of it.

Microphones & Mixers.. by BILL GIBSON

Lighting Dimmers

Often, a defective or just inherently noisy lighting dimmer causes the noise. If a dimmer is causing a noise, it is usually while the lights are being dimmed. If you spot an inexpensive wall dimmer and you're struggling to find a noise, try turning the dimmer up all the way or turning the light off.

Hire a Pro

The best solution to a persistent grounding problem is to hire a qualified and experienced system designer to rewire or to supervise the rewiring of your studio. He or she can verify that all of the available electrical outlets are properly grounded and that the electrical supply systems are suitable for powering audio and media systems.

For large audio installations, it's ideal to have the power company provide a completely separate electrical feed for all audio connections. Your qualified designer can instruct an electrician in the best scheme for providing clean AC power to all outlets designated for audio equipment power needs. Ideally, these circuits should be filtered, regulated, and relayed. When designed properly, if there is a loss of power, circuits will come back on in an order determined by a relay network.

Auxiliary Power Supply

It's also a great idea to have any computer-based gear on a power backup system. These backup systems have battery power that will continue the flow of current to your equipment if there's a momentary power loss or failure. You only need to be saved once by one of these systems to be a firm believer in their use. If the power flips off for just a second, all of your analog equipment will typically pop right back on; however, your microprocessor-controlled devices will need to reboot and might be down for a few minutes; plus, any work that you've been doing could be lost forever! Quick side note: autosave is a great feature. Use it.

An auxiliary power supply, also called an Uninterruptible Power Supply (UPS), is constantly charging its internal batteries so that when there's an AC power system failure, there is no loss of power to your equipment—the batteries keep supplying power continuously. However, this important device can only power the equipment for so long until it shuts off, too. Depending on the amount of equipment and the capabilities of the auxiliary power supply, you might have as

few as three or four minutes to shut everything down in a controlled and orderly manner, but when the batteries wear down, the power will cease. So, if you tend to leave your equipment on when you're away for a while, it's still very important to save your work and back up your files as if you were shutting everything down.

A large auxiliary power supply is not inexpensive but you need a unit capable of running your equipment long enough to close down properly. In addition, more substantial supplies also filter out spikes and surges while regulating voltage. Most home studios function well running on one 15- or 20-amp circuit so the UPS should be chosen to cover those needs. An auxiliary supply that specifies 1500–3000 watts is ample for most setups and should cost between a few hundred and several hundred dollars—you can also spend thousands for very large and capable units.

The quality and feature set of the UPS are important. Spending just a little more can provide greater amounts of power backup for a longer time, but it can also provide more tightly regulated voltage (from peaks and dips in voltage) and much greater surge and spike protection for you electrical gear, computers, modems, phone lines, and so on.

Power Conditioner and Sequencer

An auxiliary power supply performs multiple functions but should ideally be used in conjunction with a power conditioner and, when warranted, a sequencer. A good, high-quality 15-amp power conditioner provides sophisticated surge and spike suppression as well as noise filtering and rejection. A power conditioner should be used in conjunction with an auxiliary power supply to guard against outages, to maintain constant and clean power, and to protect your equipment from damaging spikes and surges.

A sequencer is often combined with the conditioner but is also frequently separate. Its purpose is to control the order of circuit/outlet power-up and power down, which helps avoid pops in the monitors from everything turning on or off at the same time. Sometimes, separate switches are available for a user-controlled shut down/power-up, but more high-quality, expensive, and sophisticated devices are microproccesor-controlled for automatic sequencing order.

Be prepared to spend another few hundred dollars to cover these needs. They're worth it, especially if you're in a problematic location with frequent surges, spikes, sags, outages, RFI, and so on.

Microphones & Mixers .... by BILL GIBSON

Circuit Tester

Most electrical stores should stock a circuit tester, designed to verify that the AC circuit from the wall has been wired correctly. If you see any deviation from standard wiring when the tester is plugged into the wall outlet, don't use the circuit.

DANGER, DANGER

Remember, when things go wrong, electrically, the human body will conduct the 20 or so amps of 110-volt alternating current (AC) to which your audio equipment is connected. Because this can be very painful or even lethal, great care needs to be taken that all safety systems are functional and in place. Be extremely careful in the operation of any electrical equipment.

To get rid of ground hums, many sound operators have gotten into the habit of lifting the AC ground, using a 99-cent adapter between the AC outlet and the power cable. This is an extremely dangerous practice and should not be done. It can result in damage to your equipment and it can be deadly. Don't do it!

Proper grounding can be the single most important factor in keeping your system safe, quiet, and buzz-free. A poorly designed system can have many hums, noises, and other unwanted sounds, and it can be dangerous.

CHAPTER TEST

1. The hot lead is:
   a. the lead that carries the signal fastest
   b. the solo that sounds best
   c. the wire that carries the actual audio signal
   d. the wire that connects to ground

2. A typical guitar cable utilizes a(n) _____ with the tip connected to the _____.
   a. shield, pickup
   b. RCA connector, signal
   c. 1/4-inch phone connector, ground
   d. 1/4-inch phone connector, hot lead

3. A typical microphone cable utilizes a(n) _____ with ___ hot lead(s) connected to _____.
   a. phone connector, one, the tip and shield
   b. XLR connector, two, pins 2 and 3
   c. TRS connector, two, the tip and sleeve
   d. XLR connector, two, pins 1 and 2

4. Balanced connections are made with:
   a. XLR connectors
   b. TRS connectors
   c. RCA phono connections
   d. both a and b

5. Professional speaker cables typically utilize:
   a. XLR connectors
   b. banana connectors
   c. Speakon connectors
   d. RCA connectors
   e. b and c

6. It is acceptable to use an RCA phono connector to adapt an XLR connector because it maintains the balanced wiring scheme.
   a. True
   b. False

7. An ohm is:
   a. the unit of resistance to the flow of electrical current used to measure impedance
   b. a ground pulse
   c. a balanced audio wave resistance
   d. All of the above

8. Most low-impedance microphones exhibit an impedance that is:
   a. between 20 and 50 ohms
   b. between 150 and 1000 ohms
   c. less than 150 ohms
   d. between 10,000 and 20,000 ohms

9. It is acceptable for a high-impedance output to connect to a low-impedance but it is not acceptable for a low-impedance output to connect to a high-impedance input.
   a. True
   b. Flase

10. It is generally accepted that unbalanced connections _____ and that balanced connections _____.
    a. should be less than 20 feet long, could be 1,000 feet long if necessary
    b. should be grounded, don't need to be grounded
    c. are susceptible to RF and other noises and interferences, effectively eliminate virtually all noises between a balanced output and a balanced input
    d. a and c

94

Microphones & Mixers.. by BILL GIBSON

11. Speaker cables utilize wire with two identical leads to provide even resistance to the positive and negative amplifier terminals.
    a. True
    b. False

12. When connecting a speaker to a power amplifier:
    a. always connect the black amplifier terminal to the red speaker terminal
    b. maintain consistent phase between the speakers and the amplifier
    c. a Speakon connector is capable of carrying up to 16 powered sends
    d. All of the above

13. Any cable with RCA phono connections at each end is acceptable to transfer SP/DIF digital audio data.
    a. True
    b. False

14. The cone speaker's physical design is very similar to:
    a. a condenser mic
    b. a moving-coil mic
    c. a ribbon mic
    d. All of the above

15. Which of these designs accentuates the bass response of the speaker cabinet?
    a. bass reflex
    b. passive radiator
    c. transmission line
    d. All of the above

16. Both high and low frequencies are equally affected by the speaker enclosure.
    a. True
    b. False

17. Combining a 4-ohm and a 16-ohm speaker in parallel results in an impedance of _____.
    a. 20 ohms
    b. 12 ohms
    c. 3.2 ohms
    d. 8 ohms

18. An active crossover changes sonic character when used for several hours in a row because it receives a powered signal from the amplifier.
    a. True
    b. False

19. When mixing down, a near-field monitor is typically the most dependable because it includes less ambient influence on the sound of the mix than a far-field monitor.
    a. True
    b. False

20. When powering down the monitor system, it is always best to turn the mixer off first so that its inputs are protected when the other gear is powered down.
    a. True
    b. False

Test answers are on page 293

Mixers

Y ou must understand everything about your mixer! Most modern mixers offer ample headroom and abundant features. The better your understanding of the primary mixer functions, the better your chances of experiencing creative freedom.

Whether you're using a digital or analog mixer—of the simplest or most complex variety—the concepts and principles of sound recording and sonic shaping hold true. With a solid foundation of knowledge about how these devices work, your question won't be, "What do I do to this sound?" It will be, "Where is the controller that lets me do what I know I want to do?"

INPUT LEVELS

Audio Example 3-1 is mixed three different ways—same music and board but different mixes. Notice the dramatic differences in the effect and feeling of these mixes. Even though they all contain the same instrumentation and orchestration, the mixer combined the available textures differently in each example.

· · · · · · · · · · · · · · · · · Audio Example 3-1
Comparison of Three Different Versions of the Same Mix

The mixer is where your songs are molded and shaped into commercially and artistically palatable commodities. If this is all news to you, there's a long and winding road ahead. We'll take

Microphones & Mixers .... by BILL GIBSON

things a step at a time, but for now you need to know what the controls on the mixer do. No two mixers are set up in exactly the same way, but the concepts involved with most mixers are essentially the same.

In this section, we'll cover those concepts and terms that relate to the signal going to and coming out of the mixing board. These concepts include:

+ High and low impedance
+ Direct boxes and why they're needed
+ Phantom power
+ Line levels

A mixer is used to combine, or mix, different sound sources. These sound sources might be:

+ On their way to the multitrack
+ On their way to effects from instruments or microphones
+ On their way from the multitrack to the monitor speakers, effects, or mixdown machine

We can control a number of variables at a number of points in the pathway from the sound source to the recorder and back. This pathway is called the signal path. Each point holds its own possibility for degrading or enhancing the audio integrity of your music.

INPUT STAGE

Let's begin at the input stage, where the mics and instruments plug into the mixer. Mic inputs come in two types: high impedance and low impedance. There's no real difference in sound quality between these two as long as each is used within its limitations.

In practical application, microphone connections are almost always low impedance, especially in the recording studio, whereas instrument outputs are typically high impedance. In order to plug a guitar or keyboard into a microphone input, you must incorporate a line-matching transformer.

The main concern when considering impedance is that high-impedance outputs connect to high-impedance inputs and low-impedance outputs connect to low-impedance inputs.

To review what we studied in the previous chapter, impedance is, by definition, the resistance to the flow of current measured in a unit called an ohm (Ω).

We put a numerical tag on impedance. High impedance has high resistance, in the range of 10,000 to 20,000 ohms. Low impedance has low resistance, in the range of 150–1,000 ohms.

A high-impedance instrument plugged into a low-impedance input is expecting to see lots of resistance to its signal flow. If the signal doesn't meet that resistance, it'll overdrive and distort the input almost immediately, no matter how low you keep the input level.

A low-impedance mic plugged into a high-impedance input meets too much resistance to its signal flow. Therefore, no matter how high you turn the input level up, there's insufficient level to obtain a proper VU reading.

OUTPUT IMPEDANCE

There is a difference between input and output impedance. In days when everything was centered on the vacuum tube (an inherently high-impedance device), it was most efficient and financially feasible to match input and output impedances. In the early 1900s, Bell Laboratories found that to achieve maximum power transfer in long distance telephone circuits, the impedances of interconnected devices should be matched. Impedance matching reduced the number of vacuum tube amplifiers needed. Since these amplifying circuits produced a lot of heat, were expensive, and were bulky, there was sufficient motivation to do whatever it took to match impedances.

Bell Laboratories invented a small cheap amplifier, called the transistor, in 1948. With the advent of the transistor (an inherently low-impedance device), everything changed. The transistor utilizes maximum voltage transfer where the destination device (called the load) should have an impedance at least 10 times that of the sending device (the source). This concept, known as bridging, is the most common circuit configuration used to connect audio devices.

Because of the load-source relationship it is possible to simply split one output several times for connection to multiple inputs. Conversely, summing multiple sending signals (sources) to one destination device (load) is not recommended. It's necessary to utilize a summing circuit to combine multiple sources to a load.

Microphones & Mixers.. by BILL GIBSON

Output Impedance Versus Input Impedance

It's alright to split outputs, typically up five or more times, to send a signal to multiple destinations. It's not okay to simply combine multiple sources to a signal input without a specific summing matrix like the output bus selector on your mixer.

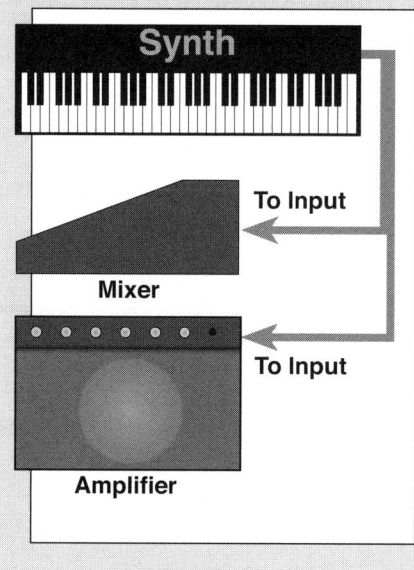

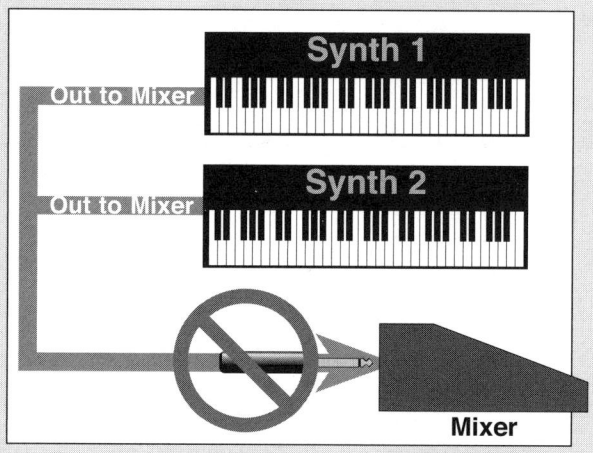

DIRECT BOX

It is standard procedure to use a direct box to match a high-impedance output to a low-impedance input or vice versa. A direct box is also called a line-matching transformer, impedance-matching transformer, impedance transformer, or a DI (direct injection). Its primary purpose is to match the output impedance of a high-impedance instrument or device to the input of a low-impedance device. Without such a device, the high-impedance, instrument-level signal that comes from a guitar, bass, keyboard, or other high-impedance electronic instrument will not sound clean and clear when connected to the mixer's XLR input. It will typically overdrive the input, even at extremely low input levels. Although the instrument's output control can usually be set very low in order to achieve a normal-looking mixer input trim setting, the sound will still tend to easily distort and will lack transparency and clarity.

Chapter 3 .... MIXERS

99

The DI is a very important device in the recording chain and the quality of different DIs varies greatly, as does their price. It's important to use high-quality, well-respected direct boxes from trusted manufacturers. Radial Engineering, Countryman, and Whirlwind offer suitable direct boxes for most applications. Instruments that contain lots of acoustical transients, such acoustic guitar, benefit most from the exceptional DIs. Radial Engineering, Countryman, and Demeter offer exceptional DIs that sound great in situations where increased detail, transparency, and signal integrity are most crucial.

Though impedance transformers work in both directions—low to high or high to low—they are best used to match high-impedance outputs to low-impedance inputs.

There are multiple benefits from the inclusion of a direct box:

- Once the signal is transformed from high to low impedance, the full benefit of the balanced wiring scheme is realized, including long cable runs without noises and RF interference.

- DIs typically contain an input pad (attenuator), which can compensate for instruments with varying output levels. If the instrument is so hot that it distorts the mixer input, simply press the pad button and the problem will likely be solved. Some DIs offer two or three attenuation levels.

- DIs contain a Thru jack that routes the signal from the instrument—connected to the DI input—to an instrument amplifier. Therefore, the instrument can be simultaneously miked at the speaker and connected to the mixer input.

- Most DIs include a switch labelled Lift, Ground, or Ground Lift. When an instrument connection to the mixer results in a loud hum along with the instrument signal, pressing this button often removes the hum. This switch should be left in the position that connects the ground until there is a problem with 60-cycle hum. Hum is most frequently an issue when the DI is connected to the mixer as well as to an amplifier via the Thru jack.

- Many DI's offer a Polarity switch that swaps pins 2 and 3 on the balanced end of the DI. This button helps compensate for incorrectly wired cables and facilitates connections between equipment that assigns pin 3 as hot.

- Some DIs contain a Merge button that combines the signals connected to the Input and Thru jacks and delivers them together as one mono signal at

the DI balanced output. This can be convenient when connecting a stereo signal to a single channel or in other infrequent applications.

- If your DI has a Speaker button, this is designed to, when pressed, allow the DI to receive a powered signal from an amplifier. Although this can produce good results, because it takes advantage of the actual amplifier sound and tone, be careful with this connection. This is the one type of connection that has the real potential to damage the gear at the receiving end of the connection along with the DI. Consult the documentation that accompanies the DI to verify the manufacturer's suggested connection procedure. Be sure to turn the output of the amplifier off when making this connection and proceed with caution.

Passive Versus Active DIs

There are two main types of direct boxes: passive and active.

- Passive direct boxes are the least expensive and generally do a fine job of matching output and input impedances. There can be a radical difference in

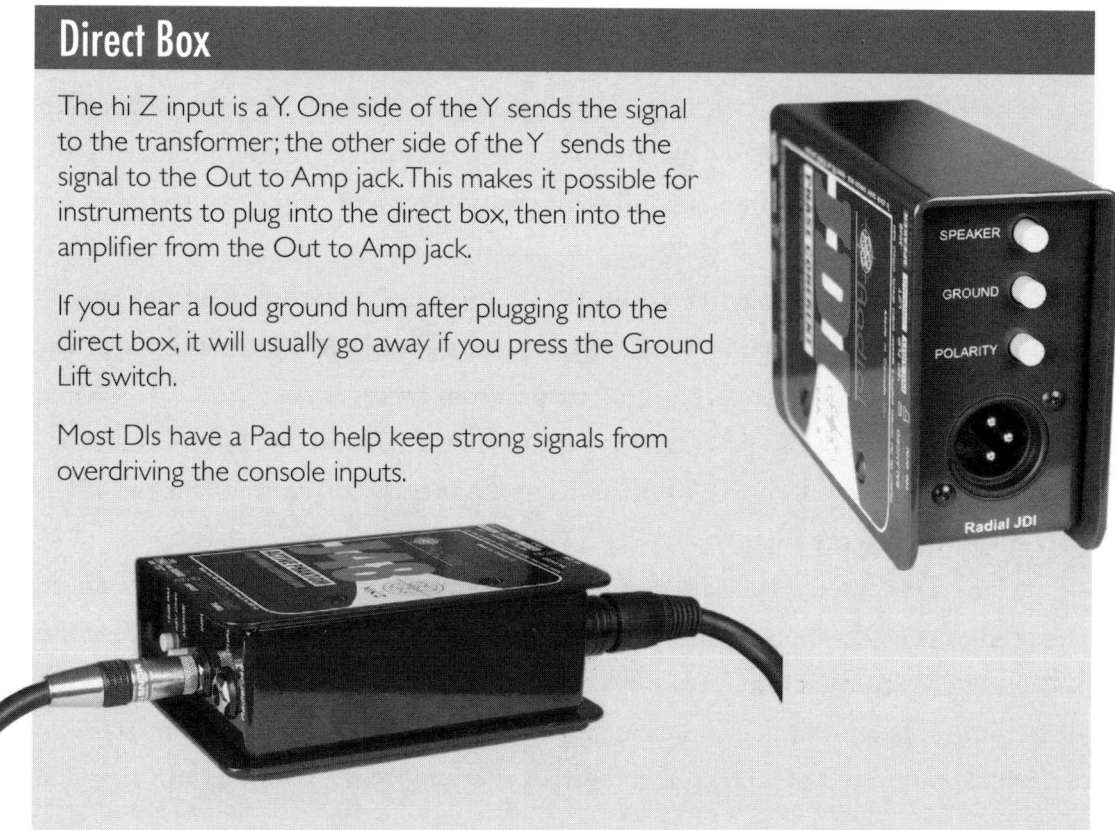

Direct Box

The hi Z input is a Y. One side of the Y sends the signal to the transformer; the other side of the Y sends the signal to the Out to Amp jack. This makes it possible for instruments to plug into the direct box, then into the amplifier from the Out to Amp jack.

If you hear a loud ground hum after plugging into the direct box, it will usually go away if you press the Ground Lift switch.

Most DIs have a Pad to help keep strong signals from overdriving the console inputs.

SPEAKER
GROUND
POLARITY
Radial JDI

Chapter 3 .... MIXERS

the sound of some instruments through a passive direct box. As with every-thing, the quality of the device affects the sound.

• Active direct boxes are usually more expensive and contain amplifying cir-cuitry that requires power from a battery or other external power supply. These amplifying circuits are used to enhance bass and treble in an attempt to regain the portion of the signal that is lost in the impedance transforma-tion. An active direct box typically gives your signal more punch and clarity in the high- and low-frequency bands.

Whether purchasing a passive or active direct box, test a few different options. As with all purchases, a good relationship with a pro audio supplier can be invalu-able when it's time to audition equipment for your specific application.

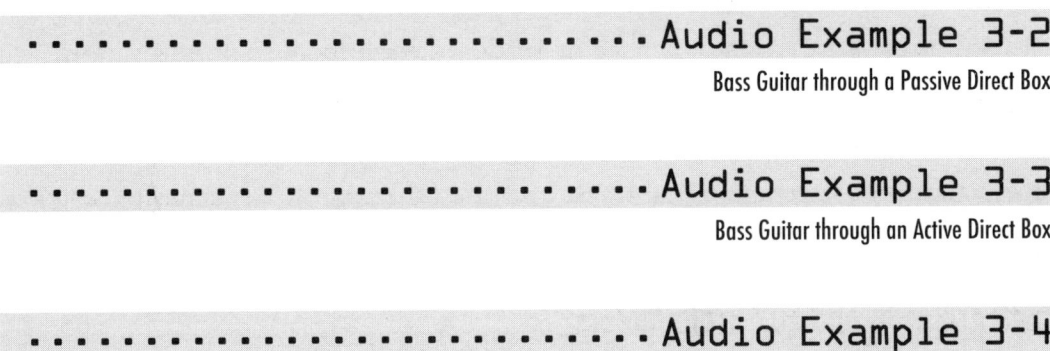

·························Audio Example 3-2
Bass Guitar through a Passive Direct Box

·························Audio Example 3-3
Bass Guitar through an Active Direct Box

·························Audio Example 3-4
Direct Box Comparisons

The difference between the previous examples can be subtle or extreme, depending on the DIs and the instruments being used. Always remember that it is frequently the nuances that make the difference between "okay" and "brilliant!" A 10-percent improvement in each track really impacts the final product, especially when recording a lot of tracks. Optimize every step of your recording process! It makes a noticeable difference.

METERS

We must use meters to tell how much signal is getting to the console, recorders, outboard gear, and so on. There are two different types of meters in common use today: the volume unit (VU) meter and the peak program meter (PPM), also called a full scale meter (FS).

Microphones & Mixers .... by BILL GIBSON

Volume Unit (VU)

VU stands for volume unit. This meter, with an average rise and fall time of about 0.3 seconds, reads average signal levels, not peak levels or fast attacks. A VU meter has a needle that moves across a scale from about −20 VU up to about +5 VU. The act of moving the needle's physical mass across the VU scale limits the potential metering speed.

In the modern sound reinforcement world, VU meters are the least common type of meter; however, they're commonly used on analog tape recorders and some mixing consoles. VU meters are still the meter of choice for some engineers. The levels they display more closely correlate with the perceived loudness of an audio signal, in contrast to peak levels, which are a very accurate measurements of signal levels from the instant they begin until they're gone.

VU meters are commonly used in outboard equipment to monitor input and output strength, as well as parameters such as gain reduction, and other dynamic controls. In these applications they are easy to read and provide an excellent visual representation of average signal levels and the other parameters they're assigned to indicate.

Peak Program and Full Scale Meters

Peak program meters (PPMs) and full scale (FS) meters are capable of accurately metering fast attacks from percussive sounds. Nearly instantaneous, sharp attacks, such as those from any metal or hard wood instrument that is struck by a hard stick or mallet, are called transients. Peak meters contain a series of small lights that turn on immediately in response to a predetermined voltage. Because there's no movement of a physical mass (such as the VU meter's needle), peak meters are ideal for accurately indicating transients.

Because the reaction of the meter to an incoming signal is electronic, rather than physical, the speed of the meter is adjustable by simply adjusting the speed of the electronic rise and release. Typically, a PPM has a potential rise time of about 10 milliseconds and a fall time potential of about four seconds.

Many peak program meters are selectable between peak and average readings. Average signal level readings are still valuable, especially considering that average signal strength offers a close correlation to the loudness of speech and music signals.

Peak meters are necessary for metering digital recording inputs because our primary goal, using a digital recorder, is to not record above a peak level with any signal. When recording digitally, always try to obtain the highest meter reading at some point during the program without activating the overload indicator. If digital levels are set too low, the full resolution of the digital recording process isn't realized. Low digital levels sound grainy and harsh; however, given the option of setting the record levels too high—risking overload—and setting the levels conservatively and not achieving full-scale level at some point, choose the latter. Especially in high bit-depth recordings, there is a little room for conservatism in level adjustments. If levels overload the input, the peaked levels clip the tops off the waveforms, which sounds bad and is difficult to satisfactorily repair.

······························· Video Example 3-1

VU and Peak Meter from the Same Sound Source

Adjusting Levels for Transients

A transient attack is the percussive attack present in all percussion instruments when one hard surface is struck with a hard stick, mallet, or beater (cymbals, tambourine, cowbell, claves, guiro, shakers, maracas, and so on). Transient attacks are also a consideration when reinforcing a piano or acoustic guitar—especially a steel string guitar played with a pick.

When metering instruments that contain transients with a standard VU meter, adjust levels so that the loudest sounds register between −9 VU and −7 VU. This approach results in much more accurate and clean percussive sound. The transient is usually about 9 VU hotter than the average level, so when the standard VU meter reads −9 VU, the input is probably seeing 0 VU. If you meter 0 VU on a transient, the input might see +9 VU!

Many VU meters contain a single LED, which is intended to indicate peak levels. This peak LED is normally just one red light that comes on when the signal is about to over-saturate tape, overdrive a circuit, or exceed maximum digital level. Along with the needle, the peak LED augments the metering accuracy of a VU meter. If the LED doesn't blink, the levels should be under control.

In the analog domain, it's usually okay for the peak LED to blink occasionally, but if it's on continuously, back off the input level until it blinks less or never.

104

Microphones & Mixers .... by BILL GIBSON

Peak and VU Meters

Zero on a peak meter registers maximum level; zero on a VU measures optimum average level, with several dB headroom before maximum level is attained. There is an art to using a VU meter, where your knowledge and insight into the operational characteristics influence level assessment. Peak meters are typically used for digital metering; they provide black and white assessment of level—you've reached overload or you haven't. Many engineers prefer to meter levels, especially mixdown levels, on both peak and VU meters. VU meters equate more closely to actual loudness. If you control peak levels and maintain strong VU levels throughout, you can be assured that your mix is strong and clean.

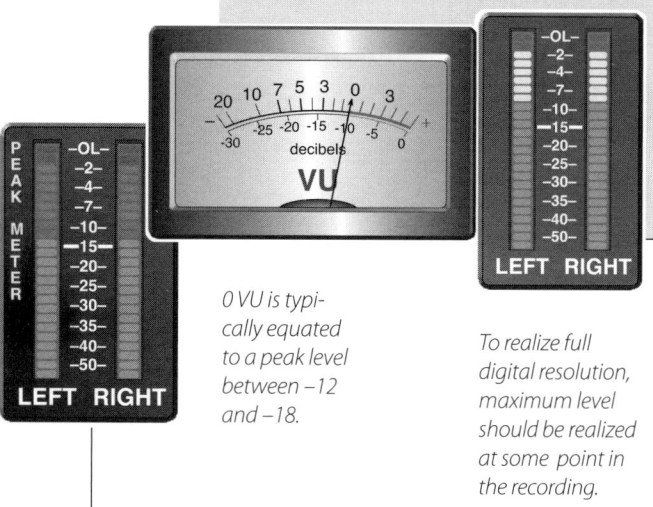

0 VU is typically equated to a peak level between −12 and −18.

To realize full digital resolution, maximum level should be realized at some point in the recording.

Typically, digital peak meters, sometimes called full-scale (FS) meters, and VU meters are compared using a 1-kHz tone. Originally, 0 VU when fed by the identical 1-kHz tone was equated to −18 on the digital peak meter. This provided plenty of headroom for transients and overloads—it also produced very conservative recording levels. As engineers began to realize that full digital level meant full vertical resolution, some manufacturers started to equate 0 VU to −12 peak—still allowing headroom for transients while providing better digital resolution. This comparison is typically viable, although when metering audio signals digitally, optimize audio resolution by achieving maximum level with no overloads.

When a peak LED comes on, it means that even though the VU is registering well within acceptable limits, the actual level that's reaching the signal path is getting pretty hot.

In a mix, if the average or VU level is conservative but the peak LEDs are always on, there's probably a percussion instrument in your mix that's too loud, and even if it doesn't sound too loud on your system, it'll probably sound too loud on other systems.

When recording most instruments, adjust the level so the VU meters read between 0 and +2 VU. For percussive instruments that have transient attacks, also called *transient peaks*, the VU meter should read around −9 VU to −7 VU at the peaks.

As a rule, meter analog signals with VU meters and meter digital signals with peak program meters.

PHANTOM POWER

Condenser microphones and active direct boxes need power to operate. If they don't receive it, they won't work. This power can come from a battery in the unit or from the phantom power supply located within the mixer.

Phantom power (a very low amperage 48-volt DC current) is available at any mic input that has a phantom power switch. Since amperage is the actual punch behind the voltage and since phantom power has a very low amperage, there's little danger that this power will cause you any physical harm, even though the power travels to the mic or direct box through the same mic cable that the musical signal travels to the mixer.

Phantom power requirements can vary from mic to mic, so check your mic specifications to ensure that the mic is getting the power it needs. Voltage requirements are typically between 12 and 52 volts. Most mics that require low voltages have a regulatory circuit to reduce higher voltages so that normal 48-volt phantom power can be used without damaging the mic. Microphones that require higher voltages won't usually sound all that great until they get the power they require. These mics often come with their own power supply.

Your mixer might not have phantom power built in. Most microphone manufacturers offer external phantom power supplies for one or more mics. Simply plug the phantom power supply into an AC outlet, and then plug the cable from the mic or direct box into the phantom power supply. Finally, patch from the XLR output of the phantom power supply into the mixer mic input.

Phantom power is preferred over battery power because it is constant and reliable, whereas batteries can wear down, lose power and cause the mic or direct box to operate below its optimum specification (even though it might still be working).

If the mic or direct box doesn't need phantom power, it's good practice to turn the power off on those channels, though it isn't absolutely essential. Many consoles have phantom power on/off switches. Some mixers have phantom power that stays on all of the time. This is okay but if there's an on/off switch, turn it on when you need it and off when you don't.

Microphones & Mixers... by BILL GIBSON

MIC LEVEL/SENSITIVITY

The effective output of each microphone is typically quantified in relation to line level, the optimal signal level for the mixer signal path and outboard equipment connections. While line level is quantified relative to a specific signal strength at a specified voltage, mic level (sensitivity rating) is quantified relative to the amount of boost its signal needs to achieve line level. Therefore, sensitivity ratings are stated in negative values, such as -20 dB or -40 dB, to indicate the distance the microphone's output level is below standard line level. The microphone output, measured in pascals, is stated relative to a specified air pressure level at the microphone capsule.

Most microphones output a signal that is between 30 and 60 dB below line level. Again, this means that the signal from the microphone must be boosted between 30 and 60 dB before the signal strength is at line level and ready to move through the mixer. In the mic specification sheet, these two possible levels would be indicated with sensitivity ratings of -30 dB and -60 dB. It's important that we understand some of the differences between various microphone sensitivities so we can see why and how the initial mixer stages are affected.

LINE AND INSTRUMENT LEVEL

Line in and line out are common terms typically associated with audio recorder inputs and outputs, mixer inputs and outputs, and outboard equipment inputs and outputs. While the signal that comes from a microphone has a strength that's called mic level, a mixer needs to have that signal amplified to what is called line level. The amplifier that brings the mic level up to line level from mic level is called the mic preamplifier—usually referred to as the mic preamp or just the mic pre. On some mixers, line-level inputs enter the board after the microphone preamp and are, therefore, not affected by its adjustment. We'll study mic preamps later in this chapter.

When comparing mic-level, instrument-level, and line-level signals, each is quantified relative to a voltage generated by a specified source output. In order of strength, mic level is the weakest; line level is the strongest. Instrument level is the most ambiguous and inconsistent of these three levels. Some instruments are capable of generating a signal that's pretty close to line level. Other instruments

Chapter 3 .... MIXERS

are extremely cold relative to line level. Some instruments, especially modern keyboards and active bass guitars, produce signals that can be patched directly into a console and function very well at line level. Most guitars connected directly into the console at line level sound weak and thin.

The Importance of Impedance Matching

It's important that we always consider both the level of the connection and the impedances of the source and input. Even if a device will functional when connected to an input—meaning you see ample level on the input meter and that it sounds okay—that doesn't necessarily mean that the connection is optimal. Guitars tend to sound pretty mediocre when patched directly into a console line input because they are high-impedance devices. They work best when connected to high-impedance inputs. Even if their level can be adjusted to function when connected to a low-impedance mixer, the sound they produce often lacks life and depth.

Instrument-level inputs are optimized to receive a range of instrument-level signals. If you're connecting an instrument and have the option, switch the input to instrument level. If there is no instrument-level input, it's usually best to connect the instrument to a mic-level input through a high-quality direct box.

Operating Levels: +4 dBm Versus –10 dBV

You might have heard the terms plus four or minus 10 (+4 or –10) used when referring to a mixer, audio recorder, or signal processor. This is another consideration for compatibility between pieces of equipment. We've already discussed the range of mic-level signals and the fact that instrument-level signals vary among various instruments. Line level connections are also divided into two commonly used signal strengths: +4 dBm and –10 dBV.

Decibel is a terms that's used in several different electronic applications. It expresses a ratio comparison between powers—a comparison of the intensity of an electrical signal to a given value on a logarithmic scale. In these references, zero on your recording meter (0 VU or typically -18 dB FS) is quantified relative the voltage it produces in response to an input of a given frequency and signal strength. An operating level of +4 dBm is specified as 1.23 volts. In response to the same input signal, -10 dBV operating level specifies .775 volts. Simply remem-

ber that +4 equipment only works well with other +4 equipment, and -10 equipment only works well with other -10 equipment.

Some units let you switch between +4 and -10, so all you do is select the level that matches your system. There are also boxes made that let you go in at one level and out at the other. In many situations, a tool like this is a necessary solution. Most +4 equipment is balanced low-impedance. This is the type of equipment that's used in professional recording studios and it uses either an XLR connector or some other type of three-pin connector, such as a 1/4-inch, tip-ring-sleeve connector. Gear can also be of the unbalanced variety and still operate at +4.

We most often think of -10 dBV gear as being unbalanced, although this is not always the case. Equipment that operates at -10 dBV is typically considered semi-professional; plus, some home recording equipment operates at -10 dBV. Gear that uses regular tip-sleeve guitar plugs or RCA phono plugs for analog connections is typically -10 dBV. Many pieces of equipment let the user select between +4 dBm or -10 dBV. Be mindful that a +4 output is too strong for a -10 input, and a -10 output is too weak for a +4 input.

When used properly, there shouldn't be a noticeable difference in sound quality from a unit operating at -10 as opposed to +4, even though +4 is the professional standard. +4 dBm balanced equipment works especially well when longer cable runs are necessary—like in a large recording studio—or when radio interference and electrostatic noises are a particular problem.

Matching Operating Levels

Use an audio level interface such as the Aphex Model 124A to connect -10 and +4 equipment together. This device matches operating level and impedance, converting the operating level in either direction: from +4 (or even +8 dBm) to -10 dBV or vice versa. A device like this allows for the use of -10 dBV consumer hi-fi equipment with +4 or +8 dBm professional/industrial/broadcast audio systems. It provides an extremely clean, reliable two-way buffer so both systems can operate at maximum performance levels, matching impedances and operating levels.

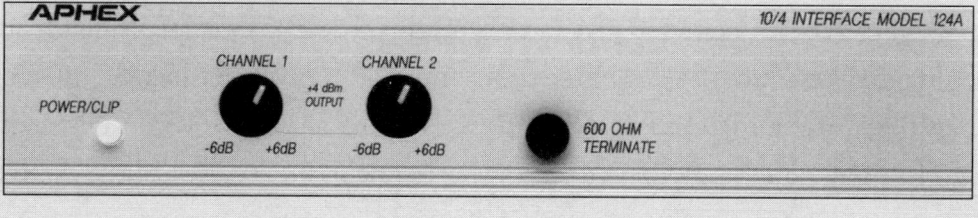

Chapter 3 .... MIXERS

ATTENUATORS

Most modern mixers have input attenuators (pads) at the beginning of each channel's signal path, between the microphone input and the mic preamplifier. The purpose of the attenuator is to compensate for extremely hot and aggressive sound sources. Instruments such as the snare drum, trumpets, and kick drum, along with loud guitar amps and other loud sound sources—especially those with substantial transient content—are very capable of overdriving the input stage of the mixer channel. Overdriving the mixer input causes distortion and negatively affects the tone, character, transparency, and impact of the sound. It's interesting to note that, since the attenuator is before the input level adjustment (gain trim), it's likely that the record level will look fine but the sound will be distorted or lifeless. When this happens, attenuation usually helps. Attenuators typically offers multiple positions, including 0, -10, -20, and so on.

Start building a sound with the attenuator in the Off or 0 position. If the overload light blinks or if the sound seems distorted—even slightly distorted—move the pad to its first level of attenuation and use the gain trim to reset the level. Some mixers have attenuators on the line inputs and the mic inputs to compensate for different instrument, recorder, and playback device output levels. As we optimize each instrument or voice recording, we must optimize the gain structure at each point of the signal path. When all the levels are correct for each mic preamp, line attenuator, fader, EQ, bus fader, etc., we can record the cleanest, most accurate signal. When one link of this chain is weak, the overall sonic integrity crashes and burns. Anything that happens at the input affects the entire signal chain. If it's not sounding great at the input, it won't sound great at the output, unless of course, you want a small and weak, somewhat-distorted and powerless sound.

110

Microphones & Mixers... by BILL GIBSON

CONNECTING A PROFESSIONAL MICROPHONE TO BUILT-IN COMPUTER INPUT

There are three key factors that must be discovered and matched between the mic and the computer's sound input:

- Signal level
- Electrical impedance
- Connector type and wiring scheme

Signal Level

The computer's built-in input, typically labeled with a small microphone, is frequently designed for use with a high-impedance microphone. Check your computer documentation for the recommended microphone type. Whereas most computer sound inputs require a minimum signal level of at least 1/100th of a volt (10 millivolts) or even more, the average professional microphone typically provides a mere 1/1,000th of a volt (1 millivolt).

There are two possible solutions to this level incompatibility:

1. Often, the computer sound card has a level boost built in, which might add sufficient gain to provide adequate level. However, these amplifying circuits often cause an unacceptable noise threshold.

2. Utilizing a microphone preamplifier is typically the best solution. Most small mixers, with microphone inputs, provide adequate preamplification of the mic signal to match the level to the computer sound card input. Plug the microphone into the mixer, and then connect the mixer output to the sound card input. By adjusting the mixer output level so that the proper levels are achieved when the sound card input is set to a normal setting, excellent audio quality is possible.

Impedance

For acceptable results, there must be an acceptable match between the microphone's output impedance and the sound card's input impedance. If the microphone impedance is substantially higher than the sound card input, some or all of the microphone signal will be lost—this is called *loading*. The higher the microphone impedance is, compared to the sound card, the more signal will be lost.

Chapter 3 .... MIXERS

111

Connectors and Wiring

The most obvious difference between the professional microphone and the computer input is the connector used on each. Simply due to space issues, most computer inputs utilize 1/8-inch mini connections. The standard XLR or 1/4-inch connectors, common to professional mics, are far too large for most computer applications.

Professional microphones with XLR connectors use a balanced wiring scheme, where pins 2 and 3 carry the audio and pin 1 and the shield are connected to ground. Most computers use an unbalanced wiring scheme, which requires only one hot lead. There is no standard wiring for computer sound card inputs, so the actual wiring scheme depends on the manufacturer.

Sound cards typically use either a mono or stereo mini plug for a connection. In the case of a mono mini plug, pin 2 of the XLR should be connected to the tip of the mini plug; XLR pin 3 and the shield should be connected to the sleeve.

In the case of a stereo mini plug, the XLR pin 2 should be connected to the mini plug tip, whereas the XLR pins 1 and 3 should be connected to the sleeve. It is unnecessary to connect anything to the mini plug ring.

If the microphone requires phantom power, the simplest solution is to use batteries in the mic itself, since most sound cards don't supply it. It's also possible to use an external phantom power supply or simply use a small mixer to provide phantom power, the proper level, and the correct impedance.

Connecting a Professional Mic to a Computer Input

This wiring scheme works well with microphones that don't require power. Whereas most mixers provide phantom power to operate condensor microphones, most computer sound cards don't. Use this method to connect the XLR from the mic to the computer's mini plug input.

Microphones & Mixers... by BILL GIBSON

MIXER CONFIGURATIONS

We begin with the mixing board. Our approach throughout this series is that the mixing board is one of the engineer's musical instruments. Always let the music lead the way through technology. Let your ears and your heart tell you what the musical sounds should be, and then use the tools of the trade to get those artistically inspired sounds.

The state of our art has radically changed over the past several years. I actually feel quite privileged to have entered the recording industry at a period now considered historic—in a day when any serious recording was done in a commercial recording studio, where the equipment used was far too expensive for the home recordist to consider, and musical and technical details could only be entrusted to experienced professionals. In those days, a mixer was a solid and stable piece of equipment. Once you understood its layout and the capabilities, you could rest assured that every day in the studio you would sit down to the same familiar console. You could learn where to tap which module when a gremlin would raise its ugly, though familiar, head. You could figure out which channel strips sounded best and which ones didn't.

I always thought of one studio, at which I was chief engineer for a number of years, as the audio equivalent of the Millennium Falcon in the first Star Wars movie. It was very good in the heat of battle but very temperamental and responsive to only Han Solo. Even though we changed all the capacitors in the console and updated all the integrated circuit chips and really tried to make everything right, I was constantly aware that there were several occasions per day where noises and other dysfunctions became so familiar that without thinking I would end up tapping in one certain corner of the control room monitor module, or re-seating a card in the tape machine, or wiggling a patch cable to remedy an otherwise debilitating glitch.

Today's mixer is often nothing more than a digital grid in which almost anything can happen and literally, in most cases, the mixer doesn't even have audio routed through it. The mixer is a control surface that controls an audio interface. The interface actually has the audio inputs and outputs. You, the user, get to set up the way you work. You control the way audio is routed, the look and feel of the user interface, even the quality and resolution of the audio signal. For the home recordist it's likely that there is no physical mixer at all. Your mixer might

Chapter 3 .... MIXERS

113

be the virtual mixer within your audio recording software. In the modern recording world, functionality is not dependent on the mixer because the functionality resides in the software and audio interface. A control surface simply provides a quick and easy means to setting up a mix.

Let's look at how typical mixers are laid out. The terms "mixer," "desk," "console," "board," "mixing desk," and "audio production console" are used interchangeably. It is very important to build a thorough understanding of basic mixer functions, whether you're using a complicated hardware or software mixer. A virtual mixer provides operational features based on, and identical in functionality to, most functions and controls found on a physical mixer. Digital consoles contain all the same controls found on the large-format analog consoles. Software often imitates and emulates the high-quality mixers, manufactured by Solid State Logic, Neve, Trident, etc. Additionally, it has become more likely that established manufacturers of outboard gear such as dynamics processors, equalizers, and effects processors are producing their own plug-ins to match both the functionality and the audio quality of their hardware products.

Classic Analog Console

This classic vintage analog Neve 8048 console sounds great! It has a sound that is warm, appealing, and musical. There's still an important place for consoles like this in the modern recording world. Most successful producers prefer to, at least, start their projects through a console like this. Solid State Logic, Trident, and API—among others—have been manufacturing fantastic consoles for decades.

Notice that each channel strip is the same—you know one and you know them all. And guess what, they're the same everyday... That is when they're all working. Classic equipment develops an attitude over the years, but once you hear the sound, you'll agree that its worth a few scratchy pots and temperamental switches.

London Bridge Studios • Seattle, WA.

Microphones & Mixers.. by BILL GIBSON

The Split Mixer

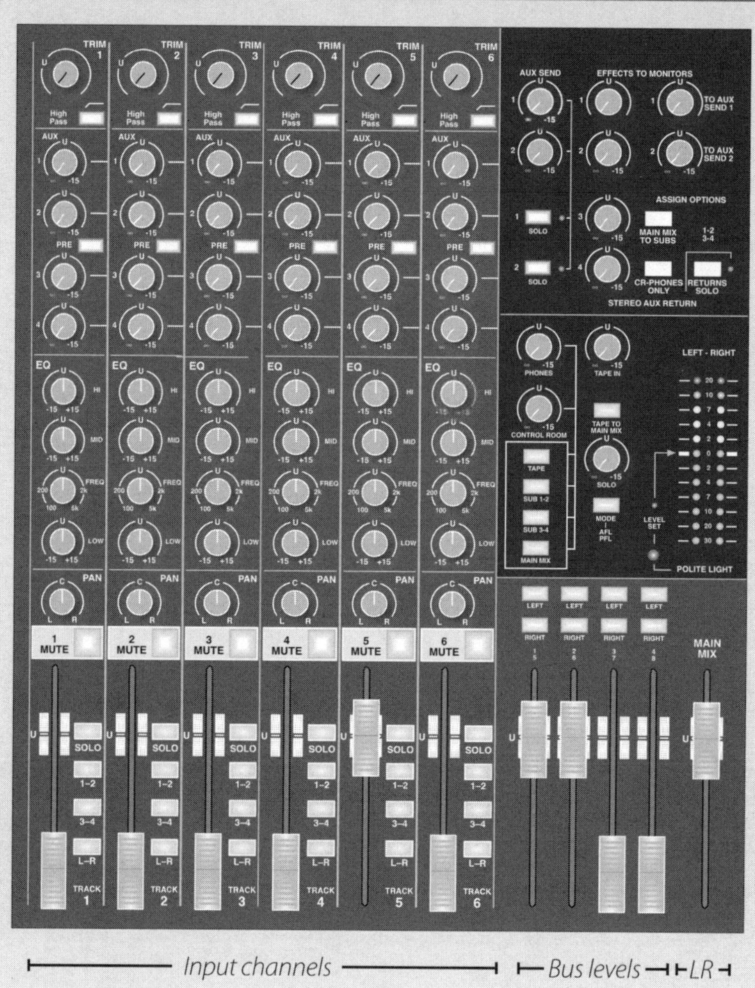

Input channels ⊢————— *Bus levels* ⊣⊢ *LR* ⊣

Mixers contain a number of channels, each typically with the same controls. These controls can include an attenuator (also called a *pad*); a phase switch; a pre-amp control; auxiliary sends; equalization; a pan control; track assignments; solo, mute and PFL buttons; and the channel fader. In understanding how one channel works, you'll understand how they all work.

To the right of the channels, there is often a monitor section. In the monitor section, there are usually a master volume control for adjusting listening levels, a monitor selector (where we choose what we listen to), master aux send levels, a test tone oscillator, a stereo master fader, a stereo/mono button, and a headphone jack. On some mixers we also see the output level controls for the track assignment bus.

The Split

The split mixer provides input channels that are routed via a switching matrix to the bus outputs. The bus outputs are typically assigned to the left-right main mix, although they can be routed to multiple locations, providing maximum flexibility. In tracking mode, the switching matrix supplies signal to the multitrack; in mixdown mode the channels are usually routed directly to the main mix unless the bus outputs are used for subgroup control.

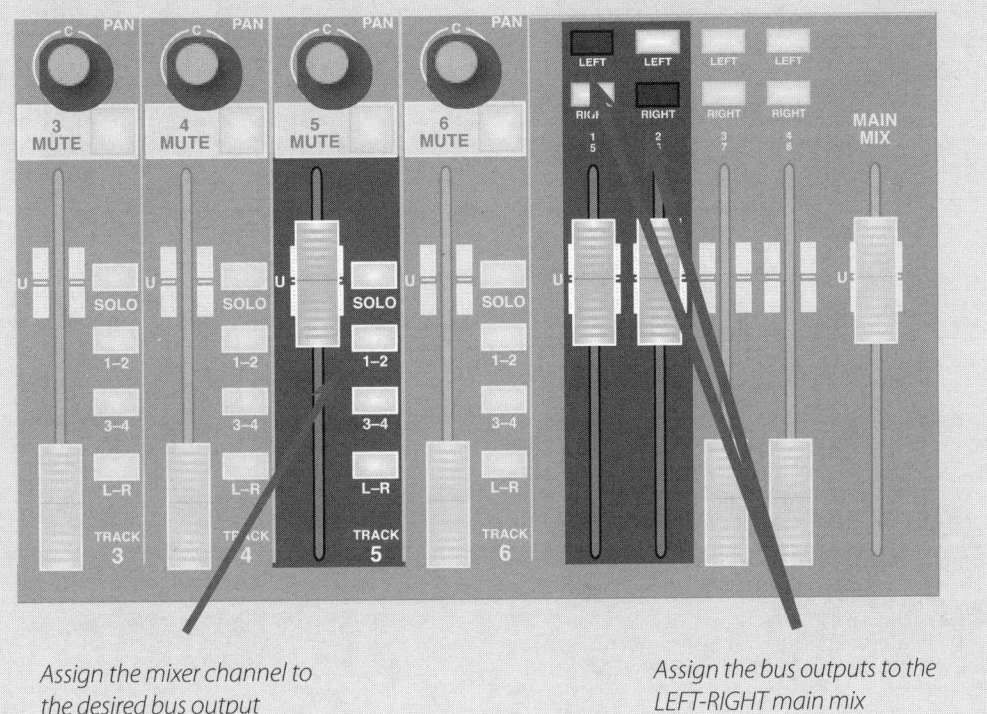

Assign the mixer channel to
the desired bus output

Assign the bus outputs to the
LEFT-RIGHT main mix

The Split and In-Line Console Designs

If your mixer has faders to the right of the channel input faders, and if these faders adjust the level of the final output to the multitrack, then your mixer is called a *split* mixer or console.

Some mixers have the level controls to the multitrack (typically knobs instead of faders) near the top of each channel. These are called *in-line* consoles.

Split and in-line mixers each have their own set of advantages, but both can be very effective and flexible while sonically supporting your musical ideas. Once you understand the conceptual design of each configuration, you'll be able to adapt quickly to either.

116

Microphones & Mixers... by BILL GIBSON

In-Line

My goal is to explain the basics in simple enough terms that you'll be able to integrate all concepts seamlessly into any recording situation. One such concept involves record bus assignments and record bus level outputs. This is an additional consideration for the recording engineer that isn't really a concern for the live engineer.

Chapter 3 .... MIXERS

During recording we always need to consider the input channel and the record matrix output separately. For example, if the electric guitar is connected to input channel 4 and then it is assigned to record bus 20:

- The input level is controlled on the channel 4 input trim.
- The record bus assignment on channel 4 is used to select record output 20.
- The output level control on channel strip 20 is used to control the actual output level to the recorder.

Whether you're using a split or in-line mixer, it's important that you keep track of input channels and record bus assignment outputs separately. Take a look at your mixer to see which kind of controls you have. Identify whether you have a split or in-line mixer.

This extra consideration is important to comprehend when in the analog domain, using an analog console, whether the console is connected to an analog or digital recording system. However, when the recording session is all "in the box," simple routing and assignments are controlled in-line. With the addition of aux buses and group faders that are assigned to record tracks, even the virtual mixer can take on the flavor of a split configuration.

In-Line Switching Matrix

The in-line console design provides the bus outputs at the top of each channel strip. These are the same controls as the bus faders to the right of the channels in the split console design. In this design the bus assignments are typically paired together—any individual button assigns to the selected bus. When odd and even buses are selected, the pan control moves the channel between odd and even (left and right).

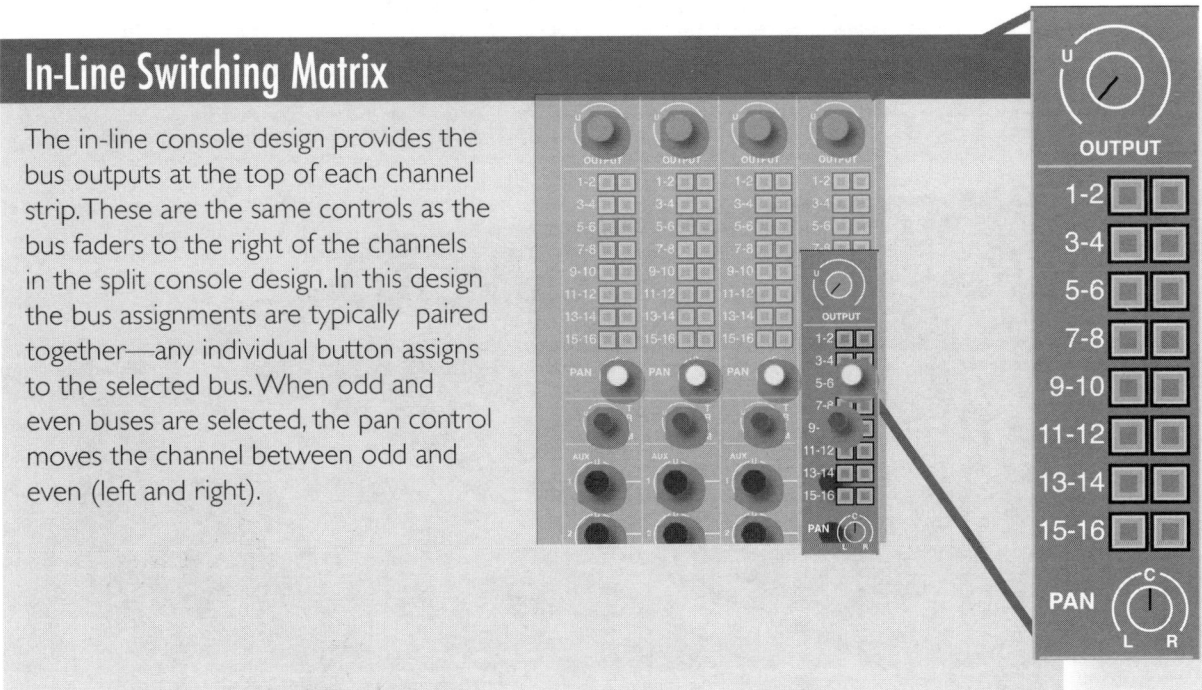

118

Microphones & Mixers.. by BILL GIBSON

Analog and Digital Mixers

Both of these consoles are powerful professional production tools. The Yamaha 02R96 digital mixer (right) provides impressive control over the audio that passes through it—much greater control than the SSL analog console (below).

Even though both consoles provide dynamics processors on each channel and fader automation, the 02R96 provides real-time automation of all dynamics, effects, levels, buses, pan, and so on. However, though they are both capable of providing a first-class professional production, the SSL is a classic in the industry. It has a distinct analog sound that is associated with many hit recordings.

With a footprint that is a fraction of the SSL's, the Yamaha offers much more audio production capacity, but its size makes it difficult for a large team to participate in its use.

In a commercial studio, a large format console such as the SSL fulfills several needs. It looks impressive, it has a high wow factor, it helps make it easy for a large team to participate in a production simply because there is more space in which to operate, and it is an indicator of the studio's commitment to excellence.

Combining Digital Mixers

New technology enables digital consoles to link together as one. These two PreSonus StudioLive 24-channel mixers connect via FireWire and function as if they were a single 48-channel mixer. They are extremely affordable and they sound great, not to mention that they provide full dynamics and effects control on each channel.

 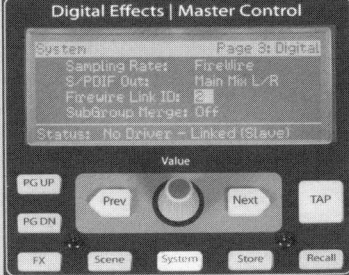

Digital mixers incorporate traditional design and layout concepts for quick and easy use. They also provide layers of control to access multiple parameters at different key locations.

Designed for either studio or live applications, these mixers can also be controlled remotely from an iPad—the interface is incredibly flexible. When held in portrait position, each channel can be accessed. Controls are readily available for compression, gating, EQ, 10 aux sends, two FX levels, and the channel mix level. Simply turn the iPad to landscape position and the screen automatically changes to one of three basic views: overview for the entire mix, aux levels for all channels and subs, or four stereo pairs of graphic equalizers that can be assigned to the mains, subs, and auxes. The Masters section is also available in this view for level and dynamics control of the mains and subs.

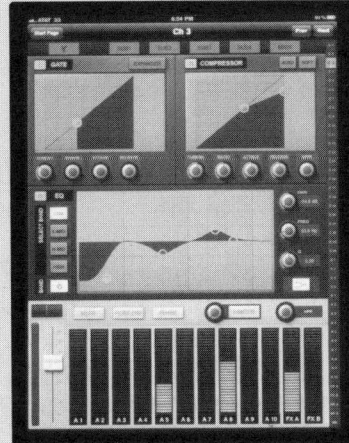 

iPad landscape and portrait views

Microphones & Mixers .... by BILL GIBSON

ANALOG VERSUS DIGITAL MIXERS

If you possess a thorough understanding of the mixer functions we've just studied, you'll be able to operate efficiently on an analog or digital mixer. With any new mixer, the real test is knowing where all the features are located—you should already have a good idea of the features you can expect to be available.

The logistical advantage of an analog console is its simplicity. Each channel is the same—the EQ knob and every other knob is at the same spot on the console each time. Digital mixers are so flexible that it's not always easy to guess where the controls are; they might be hidden somewhere in a tangled web of menus.

The really nice feature of nearly any digital mixer is its ability to store snapshots of the entire mixer layout for retrieval later. It's a simple matter to set up a tracking session and then just save the snapshot of that session. The snapshot contains all EQ, level setting, routing, effects, etc., so resetting for tracking on another day is a simple matter of a button push to recall your previously saved settings.

In addition, digital consoles provide automated control of virtually every knob, button, and parameter. When it comes to mixdown, the modern engineer can finely craft each small segment of a production a little at a time, writing the data into the automation system. Once the mix is completed and it's time to render it to the mixdown format of choice, the mix engineer simply starts playback and sits back and enjoys his or her creation.

CONTROL SURFACES

A control surface feels and acts just like a digital console. There is, however, a primary difference between the two:

- The digital console has audio inputs and outputs as well as A/D and D/A converters.
- The control surface does not pass audio signals and does not have A/D and D/A converters. The control surface has all of the faders, knobs, and switches that control all the EQ, dynamics, effects, level, pan, surround, busing, routing, inserting, sending, and so on, but must connect to the audio interface to have anything to control.

It's hard to beat the flexibility and control provided by modern control surfaces. From the small and powerful Euphonix Artist Control and Mix Control devices to the massively flexible Avid D-Command series, these devices let the engineer control all parameters of the mixer and plug-ins in many different ways.

Control Surfaces

Avid has become the dominant supplier for high-quality control, with products including:

*   *The powerful and compact Euphonix Artist and Mix Control series.*

*   *The mid-sized C|24 series.*

*   *The large format D-Control Icon series.*

Euphonix (Avid) Artist Control and Mix

Avid C|24

Avid ICON D-Command

THE VIRTUAL MIXER

A virtual mixer is the mixer built in to your digital audio recording software. Again, like the digital mixer, there aren't really any new functions on these mixers, just onscreen representations of the same functions. Sometimes the virtual mixer is more cumbersome to operate, especially if you're not using a very large monitor. However, although there aren't necessarily new functions for most basic processing tasks, the virtual mixer and plug-ins make possible many combinations and variations of parameters that are either very difficult or impossible to get from a standard hardware device.

Microphones & Mixers .... by BILL GIBSON

The Virtual Mixer

The virtual mixer is a real mixer that resides in the software application. In actuality, the modern mixer has become a combination of two components: software interface and hardware control surface. The software interface provides access to all parameters, whereas the optional hardware control surface provides a means of tactile control. The control surface provides quick and easy access to all audio parameters while feeling like a recording console.

Software Interface/Virtual Mixer

Pro Tools Mix window

The software mixer built into the modern DAW is capable of performing essentially all mixing tasks. The primary limitation is the mouse. As digital workstations have grown to dominate the home and professional studio worlds, the trend has been to provide an effective work environment, normally control-surface-based. While control surfaces are able to simulate the look and feel of a classic recording console, a patient and diligent recordist can accomplish amazing mixing feats with a mouse, some software, and a computer.

CHAPTER TEST

1. If the input level is set too high, it's alright because any problems can be fixed during mixdown.
    a. True
    b. False

2. If the input level is too low there is a good chance that the signal will contain excessive noise.
    a. True
    b. False

3. Typically, a high-impedance output connected to a low-impedance input:
    a. should work fine
    b. won't provide enough signal , so the input level will be excessively high
    c. will provide too much signal and, even if the input level is just barely up, there will still be a little distortion
    d. a and b

4. Typically, a low-impedance output connected to a high-impedance input:
    a. should work fine
    b. won't provide enough signal , so the input level will be excessively high
    c. will provide too much signal and, even if the input level is just barely up, there will still be a little distortion
    d. a and b

5. It is necessary to use _____ to combine multiple sources to a single load.
    a. a Y cable
    b. an adapter
    c. a summing circuit
    d. All of the above

6. A DI is typically used:
    a. to match high and low impedances
    b. to plug an electronic instrument into a mic input
    c. to help eliminate a ground hum
    d. All of the above

7. An active DI:
    a. has the simplest circuit and holds the potential for a very transparent sound, although inexpensive versions are often inaccurate with unacceptable signal degradation
    b. has electronic circuits designed to restore the high and low frequencies that have been robbed by the imped-ance transformer
    c. requires phantom power or batteries to power the built-in electronic circuitry
    d. b and c
    e. All of the above

8. A passive DI:
    a. has the simplest circuit and holds the potential for a very transparent sound, although inexpensive versions are often inaccurate with unacceptable signal degradation
    b. has electronic circuits designed to restore the high and low frequencies that have been robbed by the imped-ance transformer
    c. requires phantom power or batteries to power the built-in electronic circuitry
    d. b and c
    e. All of the above

9. A VU meter:
    a. has an average rise and fall time of about 0.3 seconds
    b. accurately measures transients
    c. measures average levels
    d. a and c
    e. All of the above

10. In order to realize the full digital resolution, it is best to record so that digital recording levels are as hot as pos-sible with no overloads.
    a. True
    b. False

11. When metering instruments with transients:
    a. set maximum levels at about -9 VU
    b. avoid overloads on the digital full-scale meter
    c. set maximum levels at 0 VU
    d. a and b

12. Phantom power:
    a. is typically provided by the mixer
    b. is available only at mic inputs
    c. is a 48-volt DC current
    d. travels to the device through the microphone cable or it can be supplied by a battery
    e. All of the above

13. Microphones typically output a signal that is:
    a. between 30 and 60 dB below line level
    b. between 10 and 20 dB below line level
    c. between 75 and 100 dB below line level
    d. between 5 and 10 dB below line level

14. When wiring a standard XLR mic cable to connect to the unbalanced mic input of a computer audio card, pin _____ from the microphone XLR should connect to the tip of the phone or 1/8-inch mini connector, and pins ___ and ___ should connect to the shield.
    a. 2, 1, 3
    b. 1, 2, 3
    c. 3, 1, 2
    d. 3, 2, 1

15. Many consumer-level computer sound cards contain line-level inputs only. When connecting a mic to the sound-card, the easiest solution for this problem is:
    a. to plug the mic directly into the line input
    b. to use an impedance transformer to match the mic impedance to the soundcard input
    c. to plug the mic into a small external mixer and then plug the mixer's line output into the soundcard input
    d. Either a or c

16. It is acceptable to connect a +4 output to a -10 input but not vice versa.
    a. True
    b. False

17. An in-line console has the record bus assignments at the top of each channel strip so that each channel can be easily routed to any of the recorder inputs. In addition, the output level controls on an in-line console are positioned in the channel strip, typically just below the record bus assignment buttons.
    a. True
    b. False

18. When routing a signal from the mixer input to the recorder track, the input trim is adjusted _____ and the output level to the recorder is adjusted _____.
    a. first, last
    b. for a proper reading on the mixer's input meter, for a proper reading on the recorder's input meter
    c. at the input channel, at the output control corresponding to the recorder input track
    d. a and b
    e. All of the above

19. When comparing analog and digital consoles:
    a. the analog sound is still highly touted as being warmer, smoother, and more accurate in comparison to the digital sound
    b. the automation systems on an analog console offer essentially the same flexibilities as those on a digital console
    c. the large format analog console simply looks more impressive than a similarly equipped digital console
    d. a and c
    e. All of the above

20. Many output assignment busses route the input channel to stereo pairs of outputs. A separate pan control is used to select the exact position between the two outputs.
    a. True
    b. False

Test answers are on page 293

Signal Path

CHAPTER 4

At first glance, mixers can be very intimidating to new users. Don't forget—for the most part each channel has exactly the same controls. So, if we can use and understand one channel, we've already won most of the battle. In this section, we begin to see what each control can do. As you grasp these concepts thoroughly, the mixer becomes a creative tool rather than a formidable adversary.

In the modern recording era, there are young aspiring recordists who might never use a traditional hardware mixer. To those who have come up through the analog years, that might seem pretty odd, but a new era is here. Digital Audio Workstations (DAWs) are the norm. The DAWs might be completely computer-based or they might be combined at some point with a control surface. Most new studios are built around a powerful control surface, computer software, and a high-quality interface. There is, however, still a place in the recording world for classic analog equipment. Many of the most serious and financially viable recording projects still incorporate classic analog equipment along with digital technology—most projects end up in the digital realm while still in the multitrack, pre-mixed production phase.

While some who are reading this book might not ever use an actual hardware analog mixer, the transition wouldn't be that difficult. Whether transitioning from analog to digital or digital to analog technology, the routing and signal path considerations

CHAPTER AT A GLANCE

Microphones & Mixers .... by BILL GIBSON

remain constant. In general, the stages between the source and the recorder, or the recorder and the monitors, don't change because of the technology. Though each technology has some unique considerations, the the transition between them is relatively simple.

SIGNAL PATH

A signal path is simply the route that a signal takes from point A to point B. For speed and efficiency in any recording situation, it's essential that you're completely familiar with the signal paths involved in your setup. Any good maintenance engineer knows that the only surefire way to find a problem in a system is to follow the signal path carefully from its point of origin (point A, for example, the microphone) to its destination (point B, the speakers).

There are several possible problem spots between point A and point B. A thorough knowledge and understanding of your signal path lets you deal with any of these problems as quickly as possible.

Signal Path

Many owner's manuals give a schematic diagram of the mixer's signal path. You may not be totally into reading diagrams, but there's a lot to be learned by simply following the arrows and words.

A thorough understanding of your signal path will help you troubleshoot most audio problems. Review your mixer's block diagram—follow the path from the mic to the main output. These drawings are easy to follow and, along the way, you'll see exactly how your mixer routes signals.

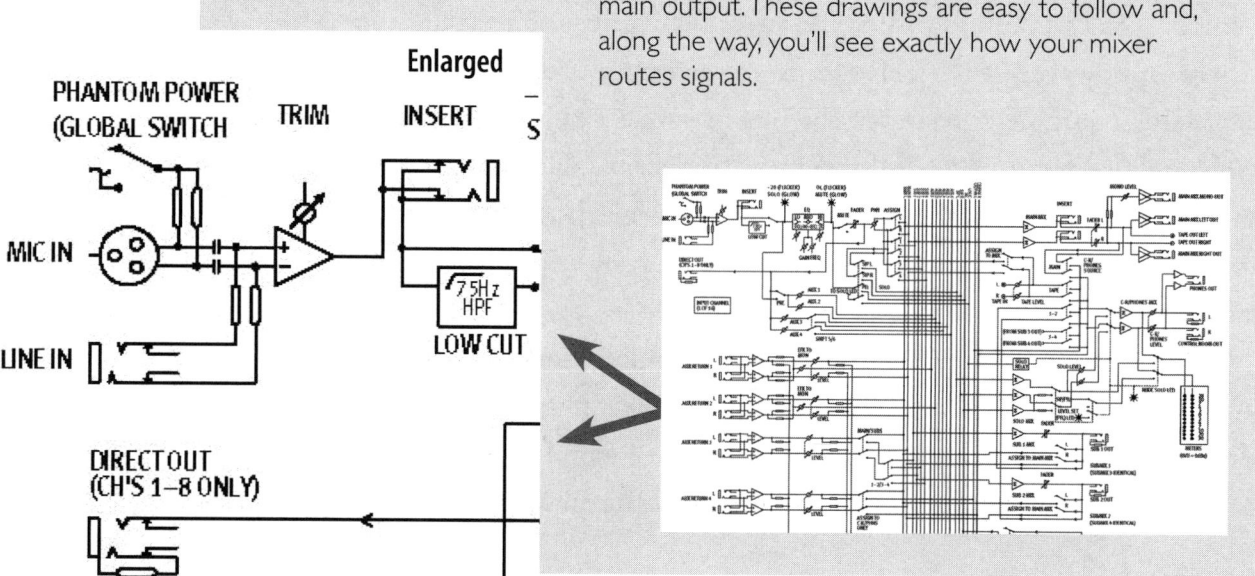

Simplified Signal Path

A thorough understanding of your signal path will help you troubleshoot most audio problems. Build a diagram like this for your recording setup. Include separate pathways for different tasks, such as:

• Recording tracks

• Mixing down

• Sending from aux buses to effects

Instrument/Voice → Microphone → Mixer Input → Preamp/Attenuator → Equalizer

Monitor Bus → Master Volume → Stereo Output → Left and Right Speakers → Left and Right Ears

Most owner's manuals give a detailed schematic diagram of the mixer's signal path. You may not feel comfortable reading diagrams, but there's a lot to be learned by simply following the arrows and words. The basis of electronics is logic. Most complex electronic tasks can be broken down into small and simple tasks. That's exactly how the recording world is. What seems like an impossible task at first isn't so bad when you realize it consists of several simple tasks performed in the right order.

An example of a typical signal path is as follows: The microphone goes into the microphone input, which goes to the attenuator, which goes into the preamp, which goes into the equalizer, which goes to the track assignment, which goes to the tape recorder, which comes back to the mixer at the monitor section, which goes to the master volume fader, which goes to the main stereo output of the mixer, which goes into the power amp, which goes to the speakers, which go to your ears, which go to your brain, which makes you laugh or cry. A thorough understanding of your signal path is the answer to most trying circumstances you'll come across.

128

Microphones & Mixers..by BILL GIBSON

Adjusting Channel Gain

There are two basic methods to adjust initial gain settings.

1. The channel on the left demonstrates the use of channel peak LED to adjust the input level. With this method, while the source is active, turn the trim up until the peak LED blinks, then back the trim off slightly.

2. The channel on the right demonstrates a unity setting on the channel fader. Set the fader to unity and then adjust the trim for the proper mix level. Most boards that use this system offer a means to meter the input signal. Many mixers display the channel level on the main left-right meter when the channel solo button is selected.

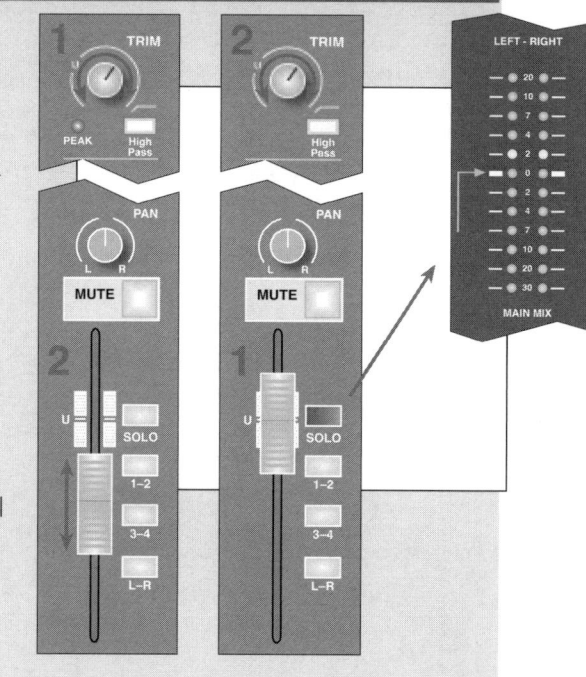

THE PREAMP

One of the first things your signal from the mic sees as it enters the mixer is the mic preamp (sometimes called the *input preamp* or simply the *mic pre*). The preamp is actually a small amplifier circuit, and its controls are generally at the top of each channel. The preamp level controls how much a source is amplified and is sometimes labeled the Mic Gain Trim, Mic Preamp, Input Preamp, Trim, Preamp, or Gain.

A signal that's been patched into a microphone input has entered the mixer before the preamp. The preamp needs to receive a signal that is at mic level. Mic level (typically 30–60 dB below line level) is what we call the strength of the signal that comes out of the mic as it hears your music. A mic level signal must be amplified to a signal strength that the mixer wants. Mixers work at line level, so a mic level signal needs to be amplified by the preamp to line level before it gets to the rest of the signal path.

The best results are usually achieved when the preamp doesn't need to be turned all the way up. A preamp circuit usually recirculates the signal back through itself to amplify. This process can add noise, and then amplify that noise,

Chapter 4 .... SIGNAL PATH

129

and then amplify that noise, etc. So use as little preamplification as possible to achieve sufficient line level.

Some mixers have an LED (light-emitting diode, usually a red light) next to the preamp control. This is a peak level indicator and is used to indicate peak signal strength that either is or is getting close to overdriving the input. The proper way to adjust the preamp control is to turn it up until the peak LED is blinking occasionally, and then decrease the preamp level slightly. It's usually okay if the peak LED blinks a few times during a recording.

Many modern mixers have very clean and transparent input preamps with ample headroom. (Headroom in any circuit, is the operational range between normal and maximum signal level.) These mixers rarely utilize a peak LED at the input stage to help set the preamp level. In this case, the channel fader and master stereo are optimized at unity gain. (Unity gain is the state where a circuit

Dynamic Range/Signal-to-Noise Ratio

When addressing sonic quality and integrity we often use two terms: dynamic range and signal-to-noise ratio. In the simplest of terms, dynamic range is the distance , in decibels, from the softest sound to the loudest sound in any audio signal. The signal-to-noise ratio (abbreviated S/N ratio or SNR) simply specifies the distance, in decibels, from the noise floor to the loudest sound in an audio signal.

Mathematically, the S/N ratio is expressed as 20 times the base-10 logarithm of the amplitude ratio, or 10 times the logarithm of the power ratio. Don't worry too much about the math for right now—simply keep in mind that we should always strive to accurately capture the full dynamic range of any source and, to accomplish this, our signal-to-noise ratio should match or exceed the dynamic range of the source.

Practically speaking, a signal with full amplitude exceeding the noise floor by 100 dB is said to have a S/N ratio of 100 dB—higher is always better. An audio source with a 100-dB maximum level and a 20-dB minimum level is said to have a dynamic range of 80 dB.

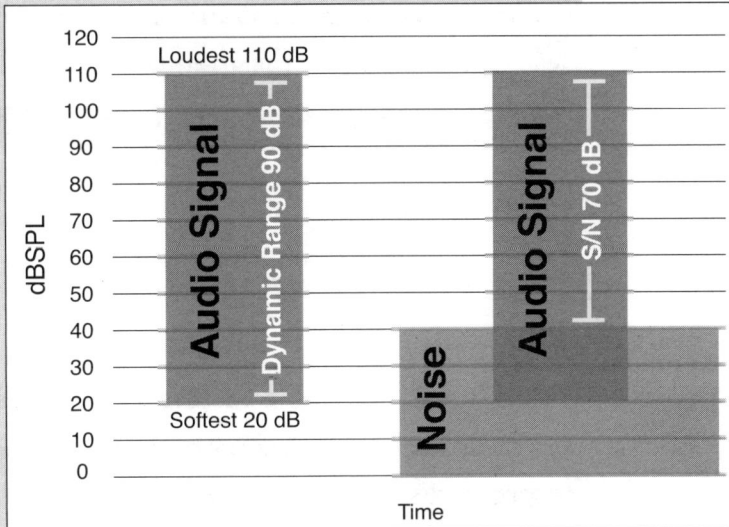

130

Microphones & Mixers.. by BILL GIBSON

outputs the same level it receives at its input.) When the channel fader is set at unity (U), the preamp level is adjusted to produce the proper level for recording or reinforcement.

ATTENUATOR

As we saw in the previous chapter, sometimes the signal that comes from a microphone or instrument into the board is too strong for the preamp stage of your mixer. This can happen when miking a very loud instrument, like a drum or electric guitar amp, or when accepting the DI of a guitar, sound module, or bass with particularly strong output levels. Some microphones produce a stronger signal than others. This is a particular problem when miking drums or loud guitar and bass amplification systems. If the signal is too strong going into the preamp, then there will be unacceptable distortion. When this happens at the input, there's no fixing it later.

This situation requires the use of the attenuator, also called a *pad*. This is almost always found at the top of each channel by the preamp level control. Sometimes, especially on condenser microphones, there is a pad between the mic capsule and the microphone circuitry. Try this attenuator before using the mixer attenuator.

An attenuator restricts the flow of signal into the preamp by a measured amount or, in some cases, by a variable amount. Most attenuators include 10-, 20-, or 30-dB pads, which are labeled −10 dB, −20 dB, or −30 dB. Listen to Audio Example 4-1 to hear the sound of an overdriven input. This example would sound clean and clear if only the attenuator switch were set correctly!

.... Video Example 4-1

Demonstration of Trim Adjustment

.... Audio Example 4-1

The Overdriven Input

If there's noticeable distortion from a sound source, even if the preamp is turned down, use the pad. Start with the least amount of pad available first. If distortion disappears, all is well. If there's still distortion, try more attenuation.

Once the distortion is gone, use the preamp level control to attain sufficient input level. Listen to Audio Example 4-2 to hear the dramatic difference this adjustment can make in the clarity of an audio signal.

.... Audio Example 4-2

Attenuator Adjustment

Again, if the input stage of your mixer has a red peak LED by the input level control, it's desirable to turn the input up until the peak LED blinks occasionally, and then back the level off until the LED stops blinking or only blinks once or twice during a take. This way we know we have the signal coming into the mixer as hot as possible without distortion. This is good in the analog domain when using a VU meter. However, in the digital domain there is no reason to push the level to peak at any time. The most important consideration is the avoidance of peaks. While many small analog mixers provide a single peak light on each

Phase Relationships When Using Multiple Mics

Whenever two or more microphones pick up the same waveform, there is a likelihood that the mics will hear the waveform at different points in the energy cycle. Notice that the two waveforms compared at the bottom of this illustration represent the portions of the waveform that were simultaneously heard at mics A and B. Even though our example uses a simple sine wave source, it's easy to see that the combination of the two waveforms will not result in the smooth and even sine wave. Considering a more complex waveform, some frequencies will combine in a destructive manner when picked up by multiple mics, and other frequencies won't.

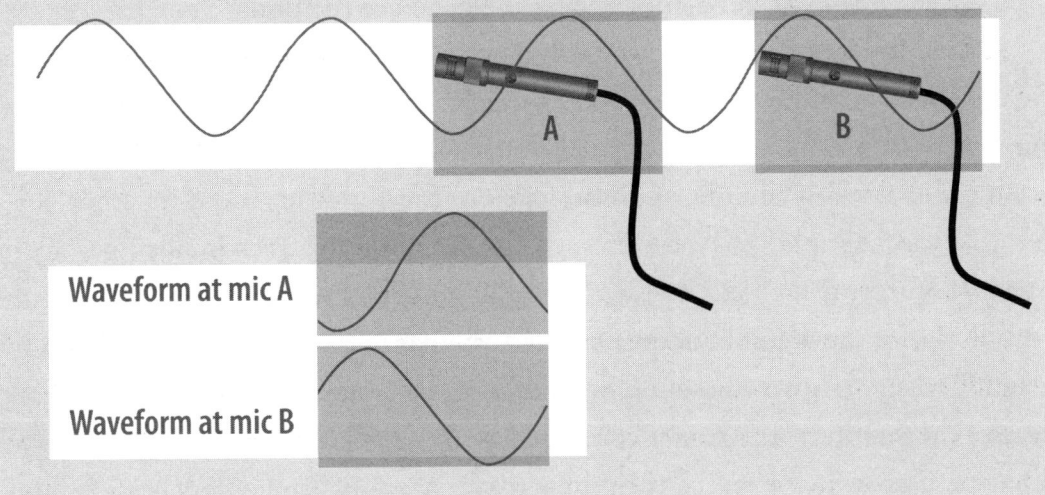

Waveform at mic A

Waveform at mic B

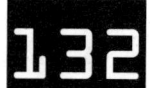

channel, any modern DAW will include an accurate onscreen meter for accurately setting audio input levels.

When possible, use the attenuator to enable the recording of electronic instruments with their output at maximum going into the board. This procedure results in the best possible signal-to-noise (S/N) ratio and provides a more sure-fire way to get the instrument back to its original level for a punch-in or retake. However, if you don't have an attenuator and if you are recording from an instrument like bass, keyboard, or guitar through a direct box, you can turn down the output of the instrument slightly to keep from overdriving the input preamp. Be sure to mark or notate the position of the instrument's volume and tone controls so you can duplicate levels for a future punch-in or retake.

PHASE

As we discovered in Chapter 1, if two signals are electronically out of phase, their waveforms are mirror images of each other. The crest of one wave happens simultaneously with the identical trough of the other wave. When this happens, there is *phase cancellation*—in other words, no signal. If two waveforms are in phase, they crest and trough together. This results in a doubling of the amount of energy from that waveform, or twice the amount of air being moved.

When phase problems occur, they're typically a result of two basic scenarios:

- An incorrectly wired microphone, or balanced line cable.
- Multiple microphones in the same acoustic space, recorded at the same time, and spaced so that undesirable phase interactions occur as the same sound arrives at the different mics in destructive phase relations.

This problem doesn't show up as much in a stereo mix, but any time your mix is played in mono or any time you are combining multiple microphones to the mix center, this can be the worst problem of all. To hear the effect of combining a sound with itself in and out of phase, listen to the guitar in Audio Example 4-3.

At the beginning of Audio Example 4-3 the original track is playing into one channel of the mixer. Next, the signal is split and run into another channel of the mixer. Notice the volume increase as the two channels are combined. Finally, the Audio Example shows the sound difference as the phase is reversed on the second track. The sound of the tracks combined is obviously thin and reduced in level. Imagine if that happened to the guitar track in a mix as it was played in mono

through your sound system at a gig, on AM radio, or through the single speaker on your smartphone or tablet computer.

····························· **Audio Example 4-3**

Phase Reversal

The nature of combining sounds dictates that there is always phase interaction. We wouldn't want to hinder that because good phase interaction gives our music depth and richness. However, we do want to be particularly aware of phase interactions that can have an adverse effect on the quality of our music.

If your mixer has a phase switch on each channel, it's probably at the top of the channel by the preamp and attenuator controls. Its purpose is to help compensate for phase interaction problems. For practical use, listen to your mixes in mono. If you notice that too many instruments get softer, disappear, or just seem to sound funny in mono, then there's probably a phase problem between some of the tracks. Change the phase of some of the tracks that might be combining in a problematic way until the mix sounds full and smooth in mono.

Short delay times, chorus, and phasing effects can also cause these kinds of problems in mono, so you might also need to change some delay times to help even things out. There will be more about this when we cover mixdown. Once you've located and solved the phase problems, your mix will sound just as good in stereo, and you'll be ready for television, AM radio, and for playback through a mono live sound reinforcement system at a show.

It's a good idea to check for phase problems when recording tracks to the multitrack. Some mixers that have phase reversal switches have them operable only on the mic inputs and not on the tape inputs. Therefore, they're unavailable during mixdown. Modern DAWs include phase inverting plug-ins so the option is always available to repair destructive phase interactions. In fact, using onscreen waveforms, it's a relatively simple matter to zoom in on problematic tracks to fine-tune the phase between tracks.

INPUT LEVEL COMPARISON

These initial variables (preamp, attenuator, meters, and phase) are very important stages in the signal path. Any good engineer has a solid conceptual grasp of them

134

Microphones & Mixers.. by BILL GIBSON

all—these are fundamental and basic considerations. You'll continually return to them for clean, quality, professional recordings.

As reinforcement of the importance of proper adjustment of the input stage of your mixer, listen to Audio Examples 4-4, 4-5, and 4-6. If your signal isn't clean and accurate at the input stage, it won't be clean and accurate anywhere.

· Audio Example 4-4
Proper Input Levels

· Audio Example 4-5
Low Input Levels Resulting in a Noisy Mix

· Audio Example 4-6
High Input Levels Causing Distortion

We must have proper levels coming into the mixer before we can even begin to set levels to tape. Any distortion or noise here is magnified at each point. Listen to the effects of improper level adjustment at the input. Audio Examples 4-4, 4-5, and 4-6 use the same song, the same mixer, and the same tracks with different input levels. All three of these audio examples are recorded at the same mixdown level; the only variable is the input trim level.

Notice in Audio Example 4-4 the sound is clean and strong. This was recorded with the input levels set properly. In Audio Example 4-5, the input level is set very low. When this happens, the levels at the end of the signal path must be elevated unnaturally in order to achieve proper mixdown levels. Along with raising levels at the final stages of the signal path comes the noise that inherently resides in the mixer circuitry.

Audio Example 4-6 is simply too hot coming into the signal path. The distortion here happens immediately. There's no way to fix the problem when the audio is distorted at the input stage.

Chapter 4 .... SIGNAL PATH

CHANNEL INSERT

Most modern mixers have what is called a channel insert. This is the point on each individual channel where a piece of outboard signal processing can be plugged into the signal path. If your mixer has inserts, they're probably directly above or below the microphone inputs.

Common Channel Inserts

A Simple Send and Return with Separate 1/4-inch jacks

Inside the mixer, the send is normally connected to the return when no plugs are in the jacks. With this sort of setup, the send and return are said to be normalled, because they are normally connected together. That connection can be interrupted by inserting a plug into one or both of the jacks.

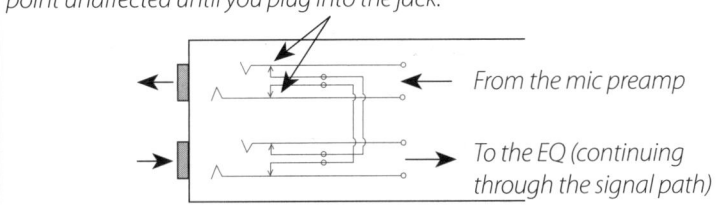

Contact at these points connects the sends to the returns when there's no jack inserted. The audio signal flows through this patch point unaffected until you plug into the jack.

From the mic preamp

To the EQ (continuing through the signal path)

The Single Insert Jack

Another common type of insert uses a single insert jack. To utilize this type of insert, you must use a special Y cable, like the one below (male tip-ring-sleeve stereo 1/4-inch phone plug to two female tip-sleeve mono 1/4-inch phone jacks).

Plug the male stereo phone plug into the insert. Next, use a line cable to connect one of the female mono connectors to the processor input, and patch the output of the processor into the other female mono connector. You might need to experiment to determine which of the mono connectors is the send and which is the return. Once everything is working, label the Y cable so it will be easy to use next time you need it.

Plugging into the jack breaks the contact here. When these points don't touch, the send and the return aren't connected through the normal.

Send

Return

From the mic preamp

To the EQ (continuing through the signal path)

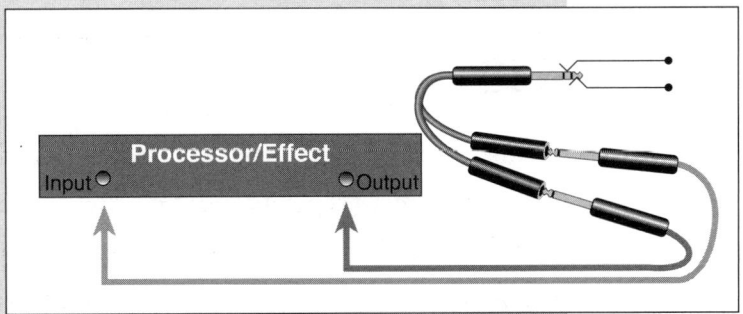

Processor/Effect

Input   Output

Channel Insert

Many mixers have a channel insert. This is the point where an outboard signal processor can be plugged into the signal path. If your mixer has inputs, they're probably directly above or below the mic inputs.

A channel insert will have a send that sends the signal, usually as it comes out of the preamp, to the processor. The output of the signal processor is then patched into the return of the channel insert. This completes the signal path. The signal then continues on its way through the EQ circuit and on through the rest of its path.

A channel insert lets you access only one channel at a time and is used to include a signal processor in the signal path of that specific channel. The processor you insert becomes a permanent part of the signal path from that point on. An insert is especially useful when using a compressor, gate, or other dynamic processor.

A channel insert utilizes a send to send the signal (usually as it comes out of the preamp) to the signal processor. The signal processor output is then patched into the return of the channel insert. This completes the signal path, and the signal typically continues on its way through the EQ circuit and on through the rest of its path.

A channel insert and an effects bus are similar in that they can both deal with signal processing. An insert affects one channel only. Inserts are ideal for patching dynamics processors, like compressors, limiters, and gates, into a signal path.

BUSES

A bus (Aux 1, Aux 2, Send 1, Send 2, and so on) lets you send a mix from the bus to an effect (typically outside the mixer but often simply routed internally to an effect, headphones, etc.), leaving the master mix on the input faders without effects. Effects buses specifically connect to a reverberation device or multi-effects processor. The output of the effect is then plugged into the effects returns or open channels on the mixer.

When we discuss the input faders as a group, we're talking about a bus. The term bus is confusing to many, but the basic concept of a bus is simple—and very important to understand. A bus usually refers to a row of faders or knobs.

If you think about a city bus, you know that it has a point of origin (one bus depot) and a destination (another depot), and you know that it picks up passengers and delivers them to their destination. That's exactly what a bus on a mixer does. For example, in mixdown the faders bus has a point of origin (the recorded tracks) and a destination (the mixdown recorder). Its passengers are the different tracks from the multitrack. Not all tracks (passengers) necessarily get on the bus, but whoever rides goes to the same destination.

Most mixers contain auxiliary (aux) buses/effects buses. Aux buses (also called *cue sends*, *effects sends*, or *monitor sends*) operate in the same way as the faders bus. An aux bus (another complete set of knobs or faders) might have its point of origin at the multitrack or the mic/line inputs. It picks up its own set of the available passengers (tracks) and takes them to their own destination (usually an effects unit or the headphones).

When a bus is used with an effect, like a reverb, delay, or multi-effects processor, the individual controls on the bus are called *effects sends* because they're sending different instruments or tracks to the effects unit on this bus. The entire bus is also called a *send*.

"Return" is a term that goes with send. While the send routes the mix of musical ingredients to the reverb or effect, the return accepts the output of the reverb or effect as it returns to the mix.

INPUT FADERS

Once everything is set properly at the preamp, use the input level faders or knobs to set the recording level. Mixing live sound is different than recording. In a live mix, the main channel faders are used to set the mix volumes for each channel. In a recording session these faders might be used during tracking to set the recording level, with another knob or fader controlling the actual monitor mix. On the other hand, there might be another knob or fader in the signal path that adjusts the level to the recorder with the main channel fader controlling the mix.

Many recording consoles provide you the flexibility to configure the layout according to your preference. Whichever way you choose to work, for our pur-

138

Microphones & Mixers .... by BILL GIBSON

Input Faders

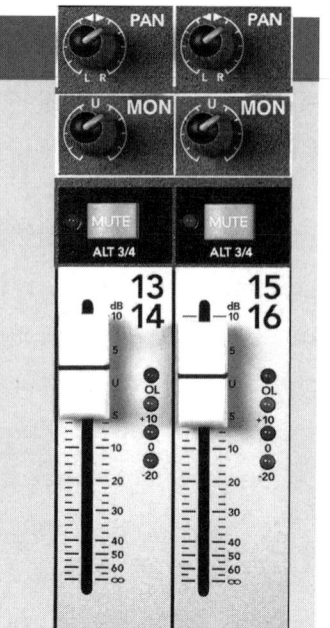

Monitor level controls receive signals from the multitrack output. When recording tracks to the multitrack, adjust the record levels with the main fader and adjust the listening volumes with the monitor level controls.

Some recording mixers offer level control via a rotary knob; some via a fader. No matter which configuration, the primary difference between a live sound console and a recording console is this concept, in which the input channel adjusts the recording level, yet the multitrack might or night not return to the same or a different channel. You must constantly keep the recorder track in mind for monitoring purposes.

poses right now, it's important that you understand this fundamental difference between a live mixer and a recording mixer. When recording, it's important to differentiate between—and keep track of—the input channel relative to the mixer playback channel. They're not always the same.

PRE AND POST

Aux buses often include a switch that chooses whether each individual point in the bus hears the signal before it gets to the EQ and fader (indicated by *Pre*) or after the EQ and fader (indicated by *Post*).

Selecting Pre lets you set up a mix that's totally separate from the input faders and EQ. This is good for headphone sends. Once the headphone mix is good for the musicians, it's best to leave it set. It's usually best if changes you make for your listening purposes don't change the musicians' mix in the phones.

Selecting Post is good for effects sends. A bus used for reverb sends works best when the send to the reverb decreases as the instrument or voice fader is turned down. Post sends are perfect for this application since the send is after the fader. As the fader is decreased, so is the send to the reverb, maintaining a constant balance between the dry and effected sounds. If a Pre send is used for reverb, the channel fader can be off, but the send to the reverb is still on. When your channel fader is down, the reverb return can still be heard, loud and clear.

Pre and Post Aux Buses

On Aux 1, the Pre-Post switch is set to Pre. This lets the Aux 1 bus hear the signal before it gets to the EQ and fader.

On Aux 2, the Pre-Post switch is set to Post. This lets the Aux 2 bus hear the signal after it has gone through the EQ and fader circuitry.

USING THE AUX BUS

Imagine there's a guitar on track 4, and it's turned up in the mix. We hear the guitar clean and dry (heard without effect). The guitar in Audio Example 4-7 is dry.

························· Audio Example 4-7
Dry Guitar

If the output of aux bus 1 is patched into a reverberation input, and the aux 1 send is turned up at channel 4, we should see a reading at the input meter of the reverb when the recorder is rolling and the track is playing. This indicates that we have a successful send to the reverb.

The reverb can't be heard until we patch the output of the reverb into either an available, unused channel of the mixer or into a dedicated effects return. If your mixer has specific effects returns, it's often helpful to think of these returns as simply extra mixer channels.

Once the effects outputs are patched into the returns, raise the return levels on the mixer to hear the reverb coming into the mix. Find the adjustment on your reverb that says wet/dry. The signal coming from the reverb should be 100 percent wet. That means it's putting out only reverberated sound and none of the dry sound. Maintain separate control of the dry track. Get the reverberated sound only from the completely wet returns. With separate wet and dry control, you can blend the sounds during mixdown to produce just

140

Microphones & Mixers.. by BILL GIBSON

the right sonic blend. This is the appropriate method whether using an analog hardware mixer or an onscreen virtual mixer. It's best to keep the dry original track separate from the 100-percent wet return. Listen to Audio Examples 4-8, 4-9, and 4-10 to hear the dry and wet sounds being blended in the mix.

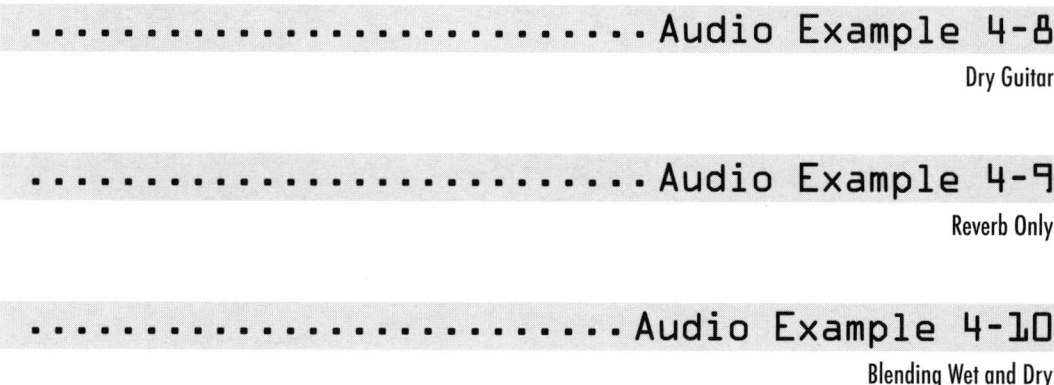

· ·Audio Example 4-8

Dry Guitar

· Audio Example 4-9

Reverb Only

· Audio Example 4-10

Blending Wet and Dry

THE HEADPHONE BUS

If your mixer has a headphone bus, or if you're using an auxiliary bus to send a signal to the headphones, patch the output of that bus into a headphone amplifier. You'll hear the mix you've sent to the headphone amp (from the headphone bus) when you plug headphones into the outputs of the headphone amp. The individual auxiliary buses on a mixer are sometimes powered just enough to run headphones; however, a good headphone amplifier provides ample power while typically offering multiple outputs with level control for recording groups of musicians.

If there's an output on your mixer labeled Headphones, it's probably powered, and you won't need a headphone amp. If you're patching this output into an amp, the powered send has the potential to overdrive the input. The resulting sound will be distorted and unsatisfactory. If the headphone output is minimally powered, you might be able to patch it into a separate headphone amp, but you must be careful to keep the headphone output level low.

On small mixers, the headphone output often derives its signal from the main faders. In some cases, there's a selector to let you listen to different buses. Large mixers and control surfaces provide extreme flexibility in the source for the headphone or aux buses.

TRACK ASSIGNMENT

The track assignment section, also called the *bus assignment section* or *switching matrix*, is used to send whatever is received at the input of the mixer (mic, instrument, or tape) to any one or a combination of output buses. These bus outputs are normally connected to the inputs of the multitrack.

A four-bus board provides you the option of sending your signal to any one or more of the four main outputs of the mixer that are connected to the multitrack recorder inputs. In the case that your multitrack recorder has more tracks than your mixer has buses, each bus is split to two or three tracks in multiples of the number of buses. For example, a four-bus mixer connected to a 16-track multitrack splits bus 1 to tracks 1, 5, 9, and 13. Bus 2 splits to tracks 2, 6, 10, and 14. Bus 3 splits to tracks 3, 7, 11, and 15. Bus 4 splits to tracks 4, 8, 12, and 16.

The bus-splitting process is especially common and frequently implemented when using four- and eight-bus mixers. Large format consoles typically have a bus output for each track on the multitrack. Consoles frequently provide 24 to 32 buses. This is very convenient when tracking a very large group, and where time is limited.

Summing

Track assignments can also be used to combine two or more outputs to one input. Patching the outputs of two or more instruments (or other devices) together through a Y cable into one input typically overdrives and distorts the input. Any time you sum (combine) multiple outputs to one input, use a circuit like the track assignment circuit on your mixer. This is designed specifically to maintain proper impedance and signal strength for its destination input. This type of circuit is also called a *combining bus, combining matrix, summing bus, summing matrix, switching matrix, track assignment bus,* or *track assignment matrix.*

The track assignment section of the mixer sums multiple channels to a single input. All or some of the channels on the mixer can be assigned to one or two tracks—that's the function of a summing bus. In addition, the track assignment matrix is equally adept at assigning a single input channel to multiple outputs. For example, if the vocal mic is connected to the mic input on channel one, the track assignment section can route it to one recorder track or to as many recorder tracks as there are output buses on the track assignment.

142

Microphones & Mixers... by BILL GIBSON

Assigning Channels to Tracks

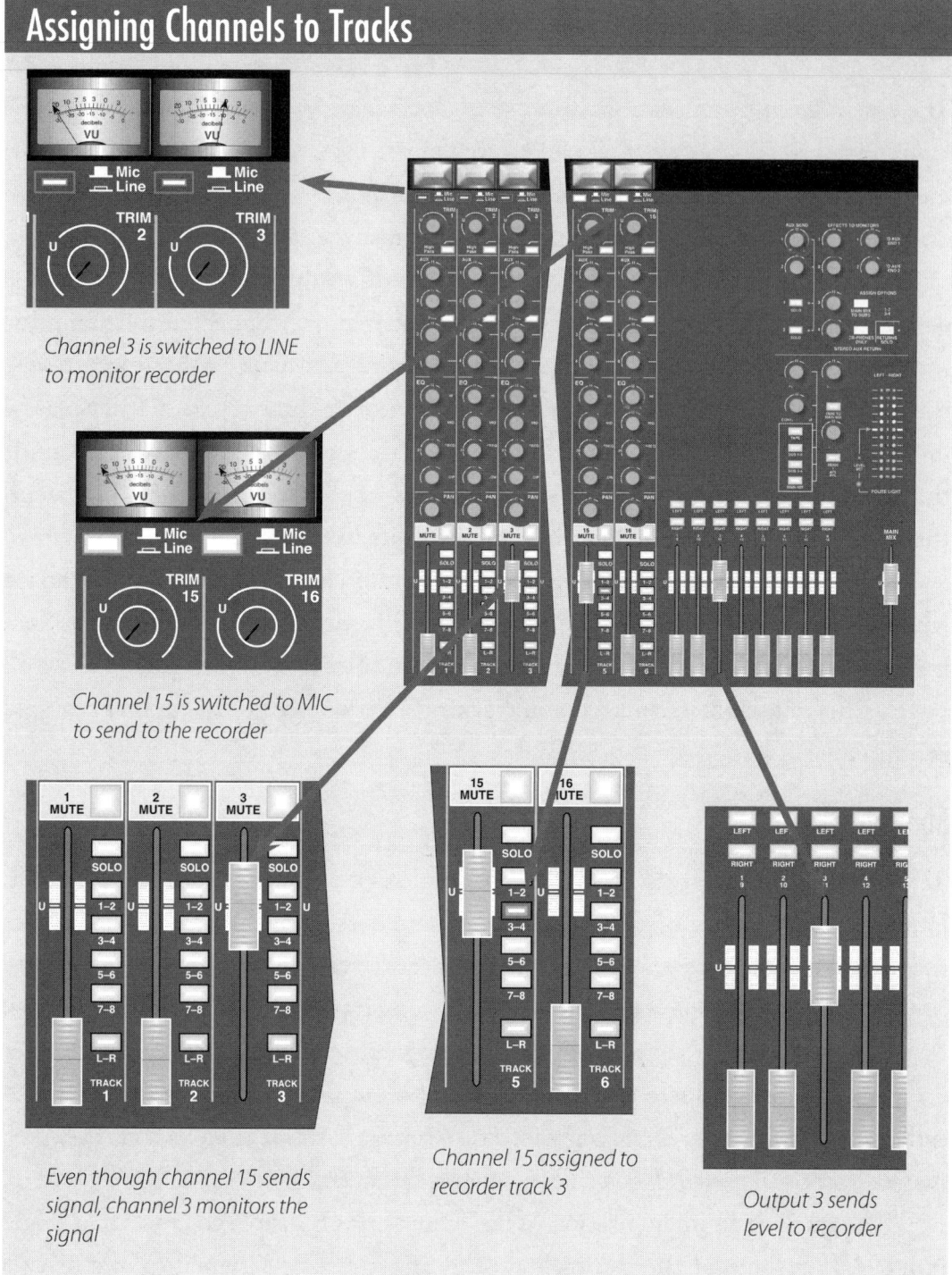

Channel 3 is switched to LINE to monitor recorder

Channel 15 is switched to MIC to send to the recorder

Even though channel 15 sends signal, channel 3 monitors the signal

Channel 15 assigned to recorder track 3

Output 3 sends level to recorder

SPLITTING A SIGNAL USING A Y CABLE

The following situations demonstrate three applications in which a Y cable could be used to split an output—a Y cable is also called a *splitter*.

1. When sending a guitar to the direct input of a mixer and simultaneously to an amplifier: You can use a Y cable out of the guitar or the guitar effects setup.

2. When sending a microphone signal to a live system and simultaneously to a recording system. In this application, a splitter box receives the signals from the stage sources and simultaneously routes them to the recording and live consoles.

3. When plugging the final output of a mixer into two or more mixdown recorders: If you're using a patch bay, all connections can be made with short, high-quality patch cords for optimum signal transfer.

Audio Examples 4-11, 4-12, and 4-13 demonstrate the sound of splitting the guitar signal with a Y cord straight out of the guitarist's effects. One side of the Y goes directly to the mixer through a direct box. The other side is sent to an amplifier. The amp is miked and the microphone is plugged into the mixer. This setup works well with a guitar, synth, drum machine, or any other electronically generated sound source. Most DIs contain a Thru jack, which is merely an internal Y that splits the incoming instrument signal through the transformer and, at the same time, back out the Thru jack.

Audio Example 4-11 demonstrates the direct guitar sound.

· Audio Example 4-11
Direct Guitar

Listen to the miked amplifier in Audio Example 4-12.

· Audio Example 4-12
The Miked Amp

We can combine the direct and miked guitar to one tape track with the track assignment bus. Listen to Audio Example 4-13 as the guitar sounds combine in different levels to create a new and interesting texture.

· Audio Example 4-13
Combining Direct and Miked Signals

144

Microphones & Mixers .... by BILL GIBSON

PAN

The pan control, sometimes called the pan pot (for panoramic potentiometer), is used to move a track in the stereo panorama. Sounds are positioned at any point in the left to right spectrum (between the left and right speakers). Some pan controls are either all the way left or all the way right with no position in between, but panning is usually infinitely sweepable from full left to full right or anywhere in between. Often the pan control is used for selecting odd or even track assignment on the multitrack bus assignments. Odd is left and even is right.

You can use the pan control along with the track assignment bus to combine multiple instruments, like several keyboards, to a stereo pair of recorder tracks. This can give you a very big sound while letting you get the most out of your equipment pool by conserving tracks and freeing up sound modules.

Listen to Audio Examples 4-14, 4-15, and 4-16. Three sounds are combined and panned around to create a unique sound that can't be gotten out of any single synth. A mixer lets you create a sound that's different from any other. Combining textures like this is called layering or sometimes doubling.

Audio Example 4-14 is synth sound A.

· Audio Example 4-14

Synth Sound A

Listen to Audio Example 4-15 to hear synth sounds B and C combined with synth sound A.

· Audio Example 4-15

Synth Sounds B and C

In Audio Example 4-16, synth sounds A, B, and C combine at different levels to become one unique and interesting sound.

· Audio Example 4-16

Synth Combination

Bouncing Analog and DTR Tracks

- Notice tracks 1, 2, and 3 are set to Tape on the Mic-Tape selector.

- Notice channels 1, 2, and 3 are assigned to tracks 7 and 8, and a mix of the three instruments is set up and panned. This mix is what will be recorded onto tracks 7 and 8.

- The output buses, 1 through 8, are connected to the recorder inputs 1 through 8. Often these outputs are split to multiple groups—1 through 8, 9 though 16, 17 through 24, etc. Notice 7 and 8 are up, sending level to the recorder.

- The recorder is monitored through channels 7 and 8. Notice that the only channels assigned to the LEFT-RIGHT bus are channels 7 and 8.

- With this setup, simply set the recorder tracks 7 and 8 to record-ready, then press play and record.

GAIN STRUCTURE

It's necessary to consider gain structure as we control different levels at different points in the signal path. Gain structure refers to the relative levels of the signal as it moves from the source to the destination. At each point where level changes are possible, you must monitor the signal strength and, if possible, solo the signal.

146

Microphones & Mixers.. by BILL GIBSON

Benefits of Bouncing Digital Workstation Tracks

Multiple benefits are provided by bouncing tracks digitally:

• All processing power being used on the bounced tracks is released once the bounce is completed. The result of the bounce includes all plug-in effects, edits, crossfades, level changes, pans, automation, etc., across all selected tracks.

• The bounce is non-destructive to the original tracks. You can always go back and re-bounce if necessary.

• A stereo bounce of many tracks is much easier to handle than several individual tracks.

We've already discussed the proper method for adjusting the input preamp level, and we've heard some examples of music recorded with the input stage too cold and too hot. These examples give an obvious demonstration of the importance of proper level adjustment at this primary stage. Each stage with user-controlled levels carries its own importance to the integrity of your signal. Ideally, you'll be able to adjust each stage to be as hot as possible, with minimal distortion.

Some mixers have a suggested unity setting for input faders and track assignment bus faders. They are usually indicated by a grayed area near the top of the fader's throw or numerically by a zero indication. Try placing the input and track assignment bus faders to their ideal settings. Then adjust the input preamp for proper recording level. This is a safe approach and works well much of the time.

Experiment with different approaches to find what works best with your setup. No approach works every time, so remember to trust your ears. If your sound is clean and punchy but the settings don't seem to be by the book, you're better off than if you have textbook settings on your mixer with substandard sound.

Confidence in your control of the gain structure can take time and experience, so start practicing. See what happens when you try a new approach.

If you adjust the input level properly, if the input fader is somewhere close to the ideal setting, and if the track assign bus/record level fader is also close to the ideal, all should be well. If one or more of these settings is abnormally high or low, you might have a problem with your gain structure.

Experiment with different approaches to find what works best with your setup. Remember to trust your ears. If your sound is clean and punchy but the

Chapter 4 .... SIGNAL PATH

147

settings don't seem to be by the book, that's better than having a textbook setting on your mixer and a substandard sound.

Potential Problems

If the input level is abnormally high (even if the signal isn't noticeably distorted), the input fader and/or track bus fader might be abnormally low. Faders work more smoothly and are easier to control in the upper part of the fader throw. When the fader is abnormally low, it's much more difficult to fade an instrument down or to fine-tune the record levels or monitor levels.

If the preamp level is destructively high, the signal will overdrive the mixer and the integrity of the entire signal path will be jeopardized. The preamp adjustment is very important. An improper setting here will result in surefire failure.

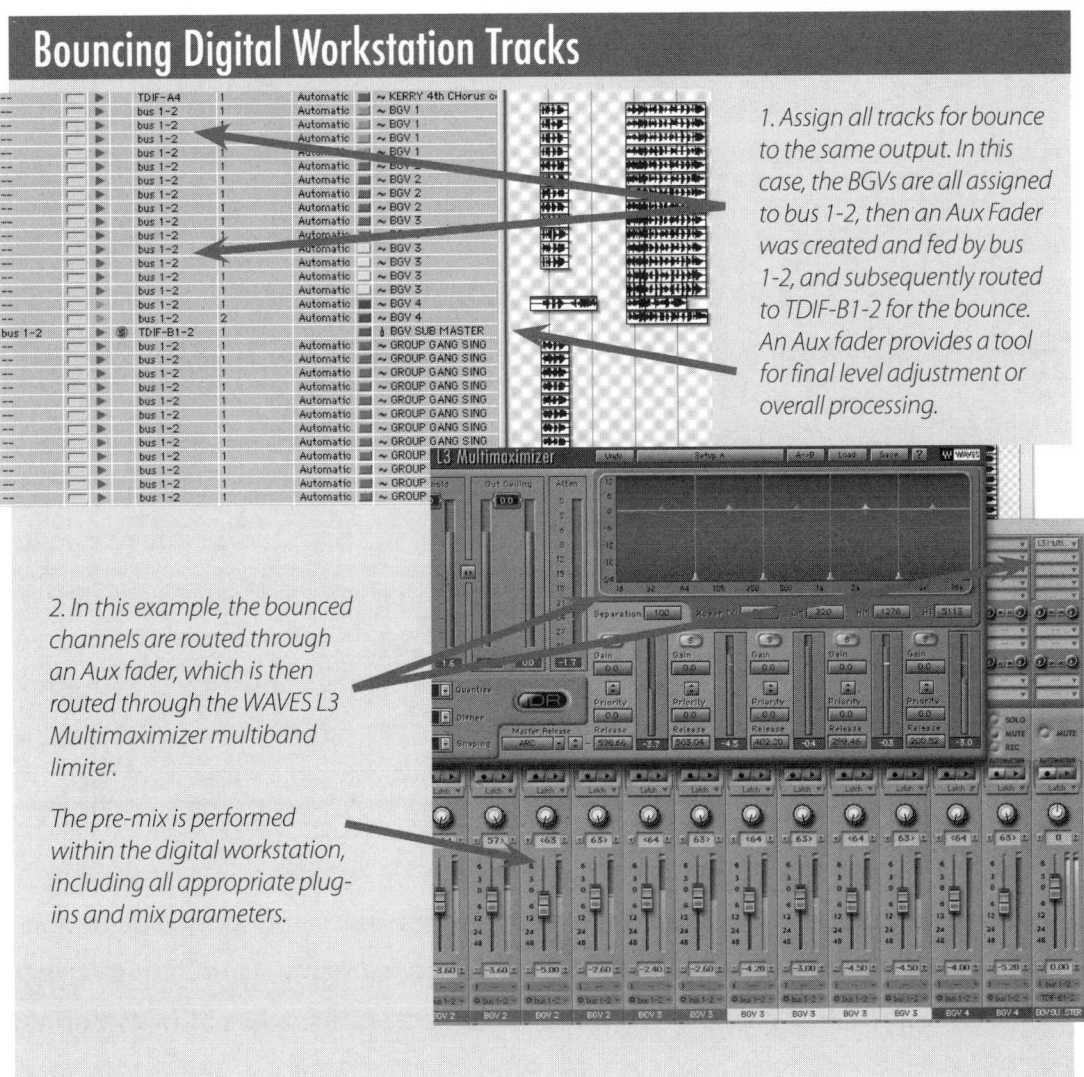

Bouncing Digital Workstation Tracks

1. Assign all tracks for bounce to the same output. In this case, the BGVs are all assigned to bus 1-2, then an Aux Fader was created and fed by bus 1-2, and subsequently routed to TDIF-B1-2 for the bounce. An Aux fader provides a tool for final level adjustment or overall processing.

2. In this example, the bounced channels are routed through an Aux fader, which is then routed through the WAVES L3 Multimaximizer multiband limiter.

The pre-mix is performed within the digital workstation, including all appropriate plug-ins and mix parameters.

148

Microphones & Mixers .... by BILL GIBSON

Bouncing Digital Workstation Tracks (cont.)

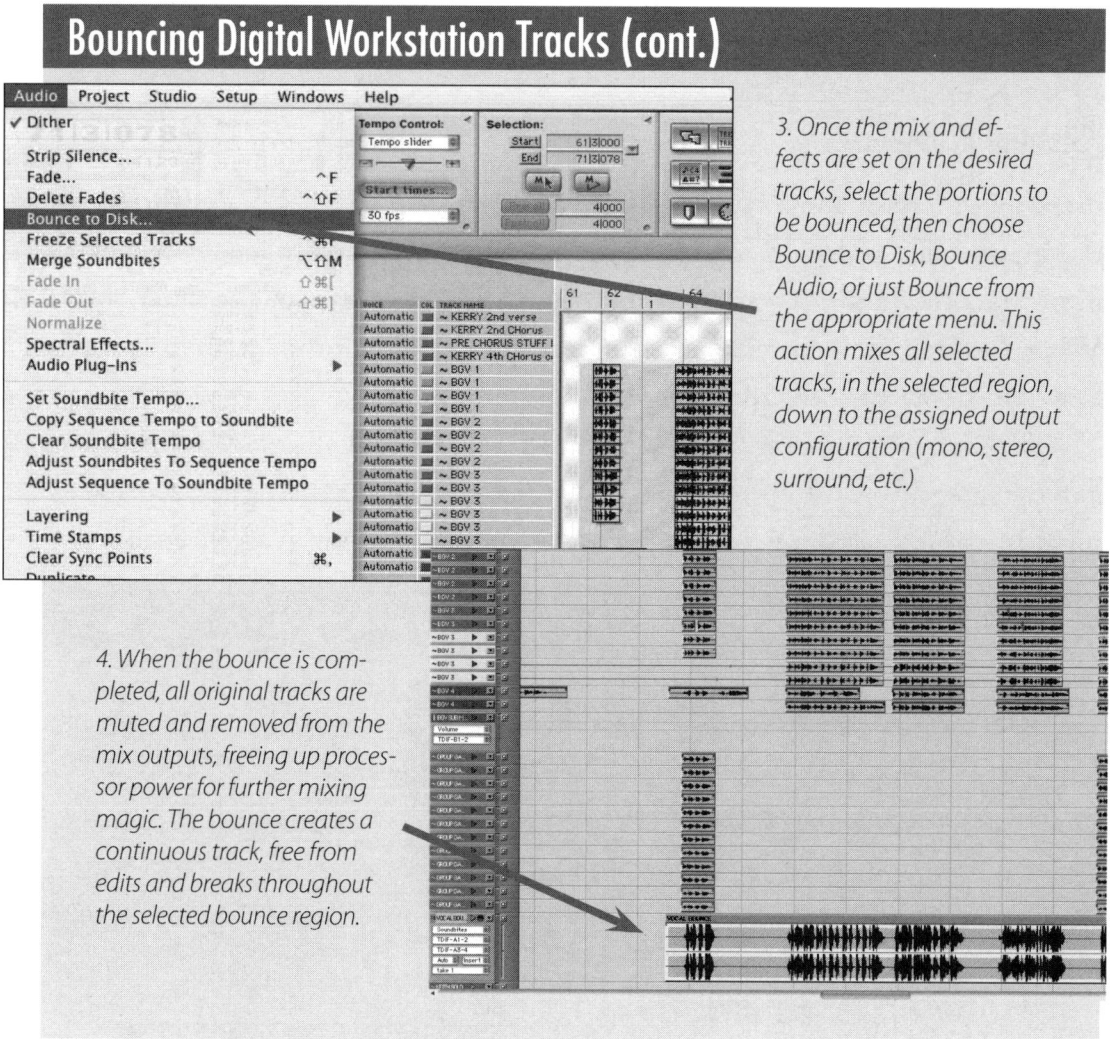

3. Once the mix and effects are set on the desired tracks, select the portions to be bounced, then choose Bounce to Disk, Bounce Audio, or just Bounce from the appropriate menu. This action mixes all selected tracks, in the selected region, down to the assigned output configuration (mono, stereo, surround, etc.)

4. When the bounce is completed, all original tracks are muted and removed from the mix outputs, freeing up processor power for further mixing magic. The bounce creates a continuous track, free from edits and breaks throughout the selected bounce region.

If the preamp level is abnormally low, the input fader and/or the track bus fader might be abnormally high. When this happens, the inherent noise that resides in your mixer is turned up further than it needs to be. Therefore, we end up with more noise in relation to signal (an undesirable signal-to-noise ratio). This is bad.

BOUNCING TRACKS

In today's recording world, there are certain topics that apply primarily to analog recording and other topics that apply to the digital realm. Combining several tracks through a summing bus to one or two other tracks, referred to as bounc-

Chapter 4 .... SIGNAL PATH

149

ing or ping-ponging, is often necessary in the analog domain in order to clear out tracks for more musical ingredients—typically backing vocals or solo overdubs.

This procedure is common in the realm of analog multitrack recording and also when incorporating the use of modular digital multitrack recorders. Given the flexibility of computer- and workstation-based multitracking in the digital domain, bouncing tracks has become less important. Using software-based recording systems, it's easy to record several tracks, and then route them all to the same output(s) for simple control. However, the bounce provides convenience in either the analog or digital domain. In the analog domain, bouncing opens up tracks so that new content can be added to the production. In the digital domain, bouncing eases the burden on the CPU, which also provides for the addition of more tracks and plug-ins without overtaxing the processer. Also, some lite versions of DAW software limit the track count substantially. In this case, bouncing is as critical as it is in the analog multitrack realm.

Analog and Digital Tape-Based Recorders

Let's take a closer look at track assignments as they're used for bouncing tracks. Since the input of a mixer can be switched to listen to mic, line, or multitrack, you select the input of two or three channels to listen to the multitrack. Once you've done this, for example, on tracks 1, 2, and 3, you can assign these channels to track 4 at the track assignment bus.

Put track 4 into Record Ready and use the faders of 1, 2, and 3 to set up the proper mix. Next, bounce those three tracks onto one by simply pressing play and record. Now start laying new parts down on 1, 2, and 3 as you listen to track 4.

If you're in the analog domain, beware of bouncing to adjacent tracks. You run the risk of internal feedback of the tape machine anytime you bounce from a track to either track directly next to it. A lot depends on the alignment of your tape machine heads and the adjustment of your playback and record electronics. Digital multitrack formats like the ADAT and 8 mm systems have no problem bouncing to adjacent tracks.

If you have the option of choosing which tracks to bounce together, the best rule of thumb is to bounce an instrument with primarily low frequencies (like a bass) with an instrument that has primarily high frequencies (such as a tambourine). This lets you adjust their relative levels by adjusting EQ. Turning down the

150

Microphones & Mixers.. by BILL GIBSON

highs turns down the tambourine; turning down the lows turns down the bass. Listen to Audio Example 4-17 to hear a demonstration of this theory.

It's also very convenient to bounce submixes of large groups of tracks, like backing vocals.

··························· Audio Example 4-17
Bouncing Multiple Instruments to One Track

Digital

Even in the digital realm, there are appropriate times for bouncing audio. If all your audio is mixed and shaped within the software package, the most efficient procedure for storing the mix is to bounce the audio, through a stereo or surround bus, to the hard drive. Most bounce options let you bounce in the format of your choice for playback on CD, DVD, etc.

There are also occasions during tracking where bouncing audio is appropriate. If you are on the verge of maxing out your computer processor, or if you're reaching the limit of your software's available tracks, simply bounce several tracks to a stereo pair. This will open more available processor power and tracks.

Since the digital bounce is typically non-destructive to the original tracks, bouncing is typically a matter of mix convenience more than anything. Many projects require several backing vocal tracks. Sometimes, for example, I've recorded 30 or more backing vocals. This becomes a logistical issue more than anything else. It is very convenient to bounce those tracks, as a stereo premix, to a pair of tracks.

Once the bounce is complete, simply turn off the original tracks, releasing processing power, and turn on the new stereo bounce. This streamlines the rest of the tracking procedure because it is much easier to control two tracks than it is to control 30. If the premix isn't just right, the original tracks are still available for revised premix at a later date. Whereas an analog bounce is destructive, because you must erase over the original tracks to record more parts, the software bounce is much less risky because the original tracks aren't destroyed as the production unfolds.

··························· Video Example 4-2
Bouncing Digital Workstation Tracks

Equalization Curve - Bandwidth

Boosting or cutting a particular frequency also boosts or cuts the frequencies nearby. If you boost 500 Hz on an equalizer, 500 Hz is the center point of a curve being boosted. Keep in mind that a substantial range of frequencies might be boosted along with the center point of the curve. The exact range of frequencies boosted is dependent upon the shape of the curve.

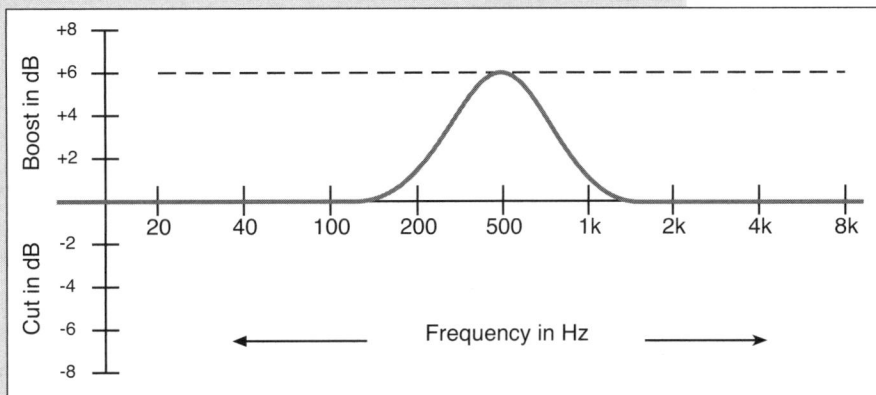

SOLO

A Solo button turns everything off except the soloed track or tracks. This lets you hear one track or instrument by itself, as if it were a solo. If multiple solo buttons are depressed, all of the soloed tracks are heard together. The solo function overrides the monitor signal. Listen as I press the Solo button on different tracks in Audio Example 4-18. You can also combine solos to hear a group of tracks together.

...........................· Audio Example 4-18

Soloing

This feature is very useful in evaluating a track for cleanliness of signal and quality of sound. It's often impossible to tell what's really going on with a track when listening to it in the context of the rest of the arrangement. There are three main types of soloing: PFL, AFL, and MIXDOWN.

PFL (Pre Fader Listen)

PFL stands for Pre Fader Listen. The PFL button solos a channel immediately before the fader. This provides an accurate picture of how a particular channel is sounding just as it's going into your mix or just before it gets to the multitrack. The position of the channel fader has no effect on the PFL solo since the signal is

Microphones & Mixers.. by BILL GIBSON

tapped prior to its arrival at the fader. Often, the PFL solo isn't affected by EQ or pan settings either.

PFL is usually the best way to verify signal integrity since it is closest to the source. If the signal is clean at the PFL position and unsatisfactory at the channel fader, a patching or console problem is likely.

The PFL solo button doesn't affect the main outputs, aux sends, or mixdown sends. Since it is non-destructive to the mix output, the PFL solo provides an excellent way to quickly verify the signal on a specific channel. In the broadcast industry, utilizing the PFL is sometimes referred to as cueing.

AFL (After Fader Listen)

The AFL solo is, as its name indicates, affected by the fader position. The signal is soloed immediately after it leaves the fader and, in addition, is typically affected by the EQ and pan and mute settings. AFL provides a convenient way to monitor a group of related tracks, by themselves, with all their level, EQ, and pan positions as they sound in the mix. A mixer with only one solo button is typically soloing AFL. AFL is sometimes called "solo in place" since it retains pan positioning.

Mixdown Solo

The mixdown solo feature is very similar to the AFL. It soloes just prior to entering the mixdown bus. All panning, EQ, inserts, and mutes are in effect, just like most AFL solo functions.

Most hardware console automation systems incorporate mixdown solo mode. With a computer-assisted mixdown, a mixdown solo button actually writes the solo into automation data. This is a very convenient feature, especially in genre where musical and textural breakdowns are common.

MUTES

A Mute button is an off button. A channel mute turns the channel off. Use the mutes instead of the faders to turn a channel down, especially when setting up a mix or setting levels for a tracking session. Beginning recordists often pull the faders down instead of using the mutes. Once you have a channel level set in relation to the other channels, you'll save time by simply muting and unmuting. The levels will remain the same, and you'll avoid continual rebalancing.

THE EQUALIZER

The equalizer or EQ section is usually located at about the center of each channel and is definitely one of the most important sections of the mixer. EQ is also called tone control; highs and lows; or highs, mids and lows. Onboard EQ typically has an in/out or bypass button. With the button set to in, your signal goes through the EQ. With the button set to out or bypass, the EQ circuitry is not in the signal path. If you're not using EQ, it is best to bypass the circuit rather than just set all of the controls to flat (no boost and no cut). Anytime you bypass a circuit, you eliminate one more possibility for coloration or distortion.

From a purist's standpoint, EQ is to be used sparingly, if at all. Before you use EQ, use the best mic choice and technique. Be sure the instrument you're miking sounds its best. Trying to mike a poorly tuned drum can be a nightmare. It's a fact that you can get wonderful sounds with just the right mic in just the right place on just the right instrument. That's the ideal.

From a practical standpoint, there are many situations where using EQ is the only way to a great sound on time and on budget. This is especially true if you don't own a wide array of mics. During mixdown, proper use of EQ is fundamental to an outstanding mix.

Proper control of each instrument's unique tone (also called its timbre) is one of the most musical uses of the mixer, so let's look more closely at equalization. There are several different types of EQ on the hundreds of different mixers available. What we want to look at are some basic principles that are common to all kinds of mixers, as well as outboard equalizers.

We use EQ for two different purposes: to get rid of (cut) part of the tone that we don't want and to enhance (boost) some part of the tone that we do want. Boosting and cutting frequency ranges are both very important. A young recordist typically reaches for equalization to add highs or lows, but rarely listens to a sound to critically locate a frequency range to cut.

Hertz

Boosting and cutting at a specified frequency number on any equalizer (for example, 100 Hz) alters more than just one frequency. It alters a frequency band that is sometimes adjustable in width. So, when we say, "Boost the bass guitar track at 100 Hz," we're really indicating a frequency range with its center at 100 Hz.

154

Microphones & Mixers .... by BILL GIBSON

The ability to hear the effect of isolating these frequency bands provides a point of reference from which to work. Try to learn the sound of each frequency band and the number of Hertz that goes with that sound.

To understand boosting or cutting a frequency, picture a curve with its center point at that frequency.

Listen to the effect that cutting and boosting certain frequencies has on Audio Examples 4-19 to 4-27.

...................................Audio Example 4-19

Boost Then a Cut at 60 Hz

...................................Audio Example 4-20

Boost Then a Cut at 120 Hz

...................................Audio Example 4-21

Boost Then a Cut at 240 Hz

...................................Audio Example 4-22

Boost Then a Cut at 500 Hz

...................................Audio Example 4-23

Boost Then a Cut at 1 kHz

...................................Audio Example 4-24

Boost Then a Cut at 2 kHz

...................................Audio Example 4-25

Boost Then a Cut at 4 kHz

...................................Audio Example 4-26

Boost Then a Cut at 8 kHz

...................................Audio Example 4-27

Boost Then a Cut at 16 kHz

Our goal in understanding and recognizing these frequencies is to be able to create sound pieces that fit together. The frequencies in Audio Examples 4-19 to 4-27 represent most of the center points for the sliders on a 10-band graphic EQ.

· Video Example 4-3

Demonstration of Equalizer Changes on Various Sounds

If the guitar track a has a balance of the entire frequency range, it might sound great all by itself. If the bass track has a very broad-range sound with lots of highs and lows, it might sound great all by itself. If the keyboard track has a huge, broadband sound, it might sound great all by itself. However, when you put these instruments together in a song, they'll probably get in each other's way and cause problems for the overall mix.

Ideally, find the frequencies that are unnecessary on each track, cut those, then enhance, or boost, the frequency ranges you like. Keep the big picture in

Frequency Ranges

The range of frequencies that the human ear can hear is roughly from 20 Hz to 20 kHz. This broad frequency range is broken down into specific groups. It's necessary to know and recognize these ranges.

- Highs - above 3.5 kHz

- Mids - between 250 Hz and 3.5 kHz

- Lows - below 250 Hz

These are often broken into more specific categories:

- Brilliance - above 6 kHz

- Presence - 3.5–6 kHz

- Upper midrange - 1.5–3.5 kHz

- Lower midrange - 250 Hz–1.5 kHz

- Bass - 60–250 Hz

- Sub-bass - below 60 Hz

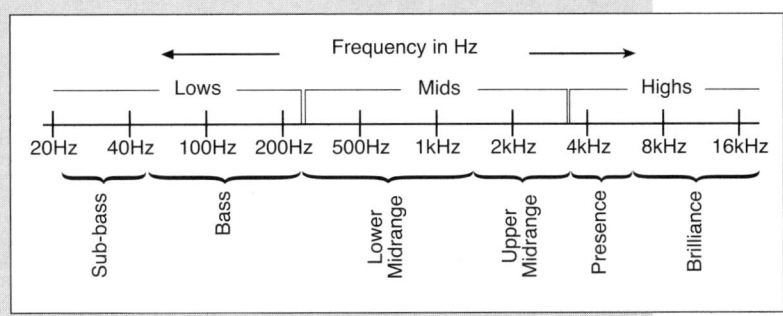

156

Microphones & Mixers .... by BILL GIBSON

mind while selecting frequencies to cut or boost. Boost and cut different frequency ranges on the different instruments and fit the pieces together like a puzzle.

For instance, if the bass sounds muddy and needs to be cleaned up by cutting at about 250 Hz and if the high end of the bass could use a little attack at about 2,500 Hz, that's great. When we EQ the electric guitar track, it's very possible that we could end up boosting the 250-Hz range to add punch. That works great because we've just filled the hole that we created in the bass EQ. Audio Example 4-28 demonstrates a bass recorded without EQ (flat).

•••••••••••••••••••••••••••••••Audio Example 4-28

Bass (Flat)

Listen to Audio Example 4-29 as I turn down a frequency with its center point at 250 Hz. It sounds much better because I've turned down the frequency range that typically clouds the sound.

•••••••••••••••••••••••••••••••Audio Example 4-29

Bass (Cut 250 Hz)

Audio Example 4-30 demonstrates a guitar recorded flat.

•••••••••••••••••••••••••••••••Audio Example 4-30

Guitar (Flat)

Bandwidth - The Q

Many equalizers let you control the width of the curve being manipulated. Notice the differing bandwidths in this illustration. Refer to bandwidth in octaves or fractions of an octave.

• Band #1 is about one octave wide.

• Band #2 is about two octaves wide.

• Band #3 is about half an octave wide.

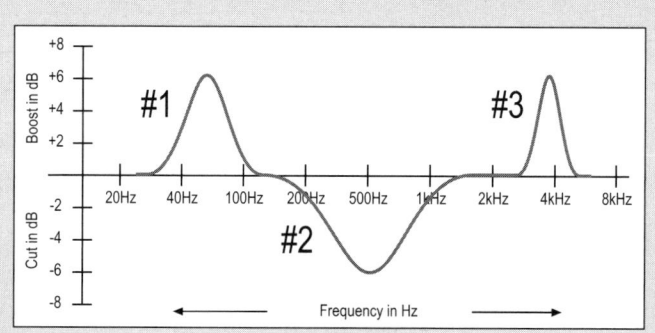

Audio Example 4-31 demonstrates the guitar with a boost at 250 Hz. This frequency is typical for adding punch to the guitar sound.

························· Audio Example 4-31

Guitar (Boost 250 Hz)

Audio Example 4-32 demonstrates the guitar and bass blending together. Notice how each part becomes more understandable as the EQ is inserted.

························ Audio Example 4-32

Guitar and Bass Together

In a mix, the lead or rhythm guitar doesn't generally need the lower frequencies below about 80 Hz. You can cut those frequencies substantially (if not completely), minimizing interference of the guitar's low end with the bass guitar.

If the guitar needs a little grind (edge, presence, etc.) in the high end, select from the 2- to 4-kHz range. Since you have already boosted 2.5 kHz on the bass guitar, the best choice is to boost 3.5 to 4 kHz on guitar. If these frequencies don't work well on the guitar, try shifting the bass high-end EQ slightly. Find different frequencies to boost on each instrument—frequencies that work well together and still sound good on the individual tracks. If you avoid equalizing each instrument at the same frequency, your song will sound smoother and it'll be easier to listen to on more systems.

Definition of Frequency Ranges

As I stated before, the range of frequencies that the human ear can hear is roughly from 20 Hz to 20 kHz. Individual response may vary, depending on age, climate, and how many rock bands the ears' owner might have heard or played in. This broad frequency range is broken down into specific groups. It's necessary for us to know and recognize these ranges.

Listen to Audio Examples 4-33 to 4-43. I'll isolate these specific ranges.

·························· Audio Example 4-33

Flat

························· Audio Example 4-34

Highs (Above 3.5 kHz)

Microphones & Mixers.. by BILL GIBSON

· Audio Example 4-35
Mids (250 Hz to 3.5 kHz)

· Audio Example 4-36
Lows (Below 250 Hz)

These are often broken down into more specific categories. Listen to each of these more specific ranges.

· Audio Example 4-37
Flat (Reference)

· Audio Example 4-38
Brilliance

· Audio Example 4-39
Presence

· Audio Example 4-40
Upper Midrange

· Audio Example 4-41
Lower Midrange

· Audio Example   4-42
Bass

· Audio Example 4-43
Sub-Bass

Some of these ranges may be more or less audible on your system, though they're recorded at the same level. Even on the best system, these won't sound equally loud because of the uneven frequency response of the human ear.

Selectable Frequencies

Two bands of EQ are available on each knob, enabling access to eight frequency bands. Pressing the Frequency Select button determines which frequency is boosted or cut. Each knob adjusts one frequency or the other, not both at the same time.

Freq. Select Button — Cut/Boost — 12 kHz / 7.5 kHz

Freq. Select Button — Cut/Boost — 4 kHz / 1.8 kHz

Freq. Select Button — Cut/Boost — 600 Hz / 300 Hz

Freq. Select Button — Cut/Boost — 150 Hz / 80 Hz

Bandwidth

Bandwidth is simply the width, quantified in octaves, of a frequency spectrum. A human being hears a bandwidth of about 10 octaves—from 20 Hz to 20 kHz.

Most equalizers contain controls for at least three bands, with each band about one octave wide. This means that the boost or cut is centered on the defining frequency but contains frequencies that extend 1/2 octave below the center point and 1/2 octave above the center point. A one-octave bandwidth is specific enough to enable us to get the job done but not so specific that we might create more problems than we eliminate. Bandwidth is sometimes referred to as the Q.

As the frequency band is raised or lowered, the frequencies on either side follow along in the overall shape of a bell curve centered on the given frequency. Bandwidth has to do with pinpointing how much of the frequency spectrum is being adjusted. A parametric equalizer is unique in that it has a bandwidth control.

A wide bandwidth (two or more octaves) is good for overall tone coloring. A narrow bandwidth (less than half an octave) is good for finding and fixing a problem.

Sweepable EQ

A lot of mixers have sweepable EQ (also called semi parametric EQ). Sweepable EQ dramatically increases the flexibility of sound shaping. There are two controls per sweepable band:

1. A cut/boost control to turn the selected frequency up or down
2. A frequency selector that lets you sweep a certain range of frequencies

Microphones & Mixers.. by BILL GIBSON

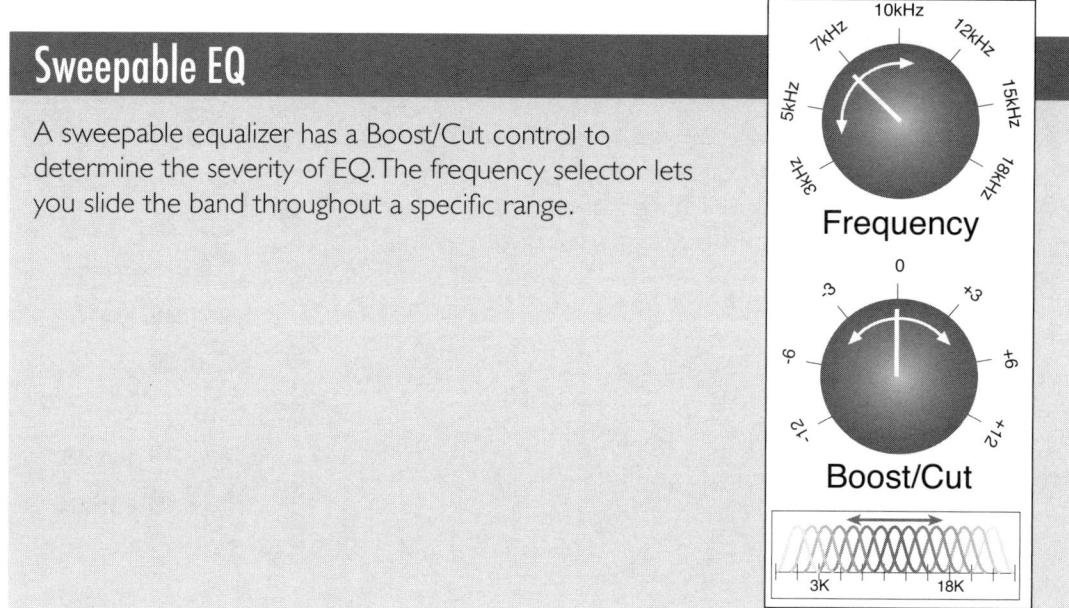

Sweepable EQ

A sweepable equalizer has a Boost/Cut control to determine the severity of EQ. The frequency selector lets you slide the band throughout a specific range.

This is a very convenient and flexible EQ. With the frequency selector, you can zero in on the exact frequency you need to cut or boost. Often, the kick drum has one sweet spot where the lows are warm and rich or the attack on the guitar is at a very specific frequency. With sweepable EQ, you can set up a boost or cut, then dial in the frequency that breathes life into your music.

Mixers that utilize sweepable EQ typically have three separate bands on each channel: one for highs, one for mids, and one for lows. Sometimes the highs and lows are fixed-frequency equalizers but the mids are sweepable.

Parametric EQ

This is the most flexible type of EQ. It operates just like a sweepable EQ but gives you one other control: the bandwidth, or Q.

With the bandwidth control, you choose whether you're cutting or boosting a large range of frequencies or a very specific range of frequencies. For example, you might boost a four-octave band centered at 1,000 Hz, or you might cut a very narrow band of frequencies, a quarter of an octave wide, centered at 1,000 Hz.

With a tool like this, you can create sonic pieces that fit together like a glove. Parametric equalizers are a great addition to any home studio. They are readily available in outboard configurations, and some of the more expensive consoles even have built-in parametric equalization.

. . . . .Video Example 4-4

Demonstration of Parametric Q and Sweep

Chapter 4 .... SIGNAL PATH

161

Parametric EQ

The width of the selected frequency band is controlled by the Q adjustment (also called bandwidth). Curve A (below) is a very broad tone control. Curve B is a very specific pinpoint boost. The Q varies infinitely from its widest bandwidth to its narrowest. Frequency and Boost/Cut operate like the sweepable EQ.

Parametric equalization is the most flexible and powerful tone control.

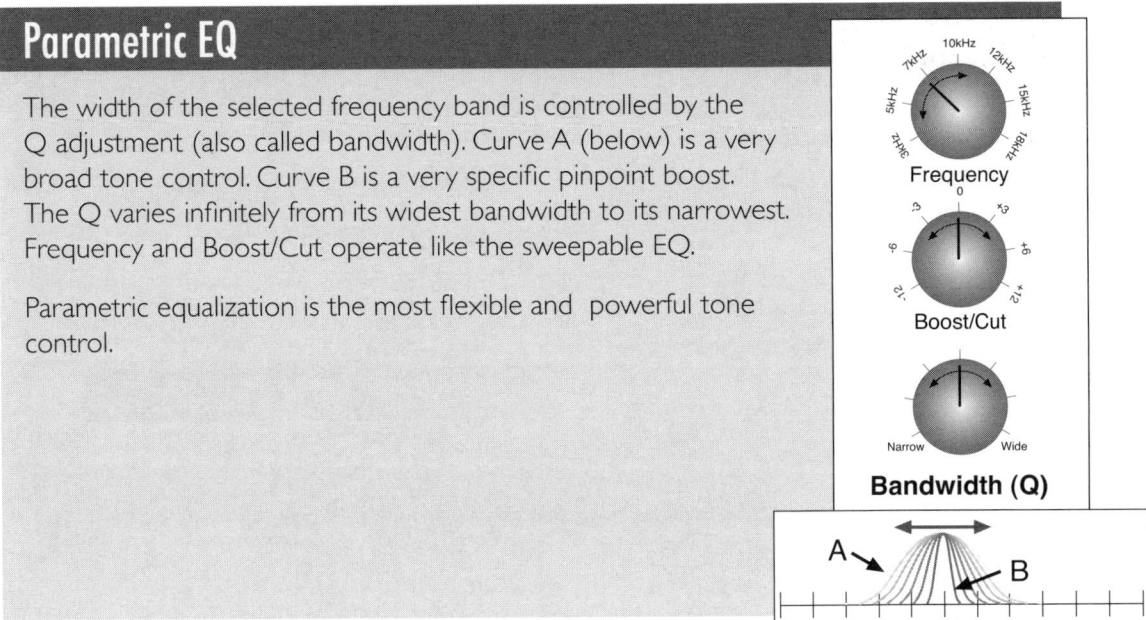

Graphic EQ

This is called a graphic equalizer because it's the most visually graphic of all EQs. It's obvious, at a glance, which frequencies you've boosted or cut.

A graphic equalizer isn't appropriate to include in the channels of a mixer, simply because of the space required to contain 10 or more sliders, but it is a standard type of outboard EQ. The graphic EQs that we use in recording have 10, 31, or sometimes 15 individual sliders that each cut or boost a set frequency with a set bandwidth. The bandwidth on a 10-band graphic is one octave. The bandwidth on a 31-band graphic is one third of an octave.

Graphic equalizers were very popular in the studio a number of years ago. Today, graphic equalizers are used mostly in live sound reinforcement applications because they are convenient and very visual, and they work well in conjunction with acoustical measurement devices.

Notch Filter

A notch filter is used to seek and destroy problem frequencies, like a high-end squeal, ground hum, or possibly a noise from a heater, fan, or camera.

Notch filters have a very narrow bandwidth and are often sweepable. These filters generally cut only.

162

Microphones & Mixers .... by BILL GIBSON

Graphic Equalizers

The 10-band graphic EQ provides good general sonic shaping. Each slider controls a one octave bandwidth.

A 31-band graphic equalizer provides control over a third of an octave with each slider.

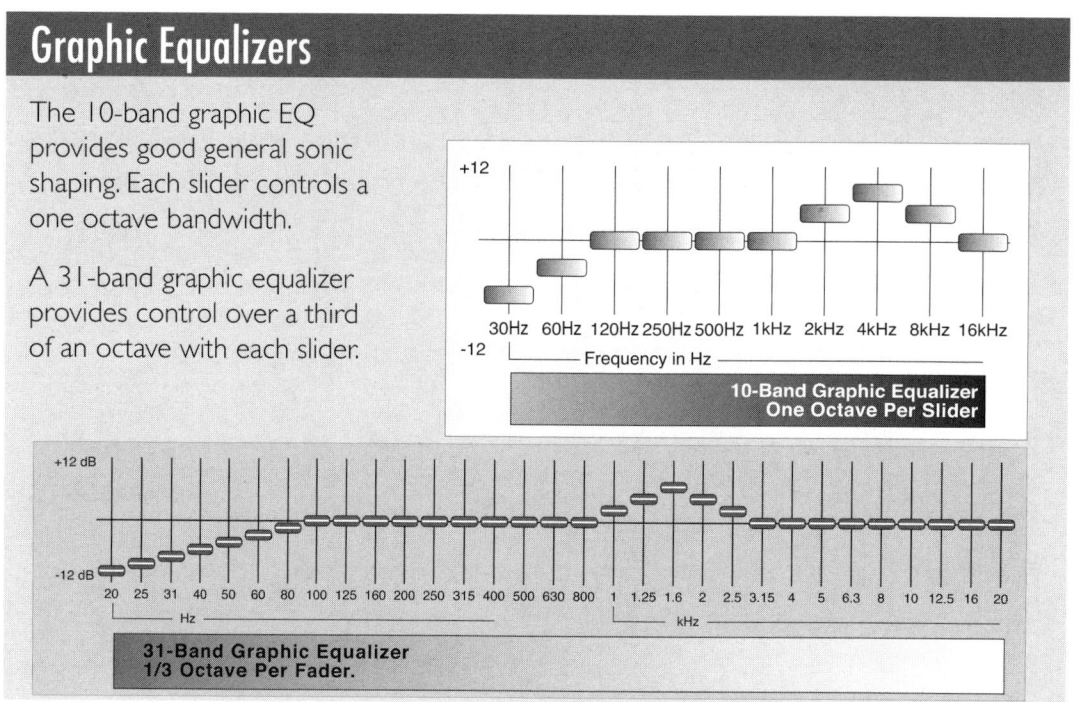

10-Band Graphic Equalizer
One Octave Per Slider

31-Band Graphic Equalizer
1/3 Octave Per Fader.

Peaking Filters

All the equalizers we've covered so far are peaking filters because they cut or boost a band in a bell-shaped curve to a peak that is centered on the defining frequency. These are, by far, the most common types of equalizers.

Highpass Filter

A highpass filter lets the high frequencies pass through unaffected but cuts the low frequencies. In previous study of equalizers, we used in image of a bell curve with a center point at the selected frequency moving up or down as the frequency range was boosted or cut. Bandpass, highpass, and lowpass filters don't fit that picture. With these filters, we specify a frequency at which the cut begins. The selected frequency is called the cutoff frequency, or sometimes the knee. Once the filter point is defined, the severity of the cut (the steepness of the filter) is calibrated in dB per octave. This rate of the cut is called the slope. In their normal use, these filters cut at a rate between 6 and 12 dB per octave.

A highpass filter can help minimize 60-cycle hum on a particular track by filtering, or turning down, the fundamental frequency of the hum. Highpass filters function very well when you need to eliminate an ambient rumble, like a furnace in the background or street noise that leaks into a vocal mic.

Most modern highpass filters provide a sweepable frequency selector. In the context of creating a mix, a highpass filter is traditionally used to trim away unused or unnecessary low-frequency information. Simply listen to the track and sweep the highpass filter upward from the low-frequency range until you hear the low end thin out a little. Using filters in this manner is a good way to clean out unnecessary sonic ingredients. However, always be careful to critically assess the sonic impact on the overall mix. This is music, after all, and sometimes ingredients we don't consciously hear are indeed affecting the emotional or physical impact of the overall sound.

Lowpass Filter

A lowpass filter lets the low frequencies pass through unaffected and cuts the highs, usually above about 8 to 10 kHz. Lowpass filters have many uses. For instance, they can help minimize cymbal leakage onto the tom tracks, filter out a high buzz in a guitar amp, minimize tape hiss, or filter out string noise on a bass guitar track.

Listen to a track with the lowpass filter set with cutoff frequency as high as possible, then lower the cutoff frequency until you can hear the highs diminish. Then, raise the cutoff frequency slightly for a natural sound, while filtering out extraneous high frequencies. As with the highpass filter, always be careful to critically assess the sonic impact on the overall mix. You might be filtering frequencies that, even though you don't think you can hear them, are combining in the mix to create a sound or a feeling.

These filters, whether high-, low-, or bandpass, should be used to filter specific unwanted mix ingredients; they shouldn't be used to trim away at every track on the mix, just in case there's a problem in a frequency range.

Bandpass Filter

A bandpass filter lets us select a frequency range (a band) and let it pass through unaffected. In other words, all frequencies above and below a specified frequency range are filtered out. The bandpass filter is just like the marriage of a highpass and a lowpass filter. With a bandpass filter, it's easy to create a lo-fi sound like that projected from a small transistor radio, or to zero in on any specific frequency range for a special effect.

164

Microphones & Mixers.. by BILL GIBSON

Types of Filters

Equalization comes in many forms. The first types of equalizers we covered were peaking filters, where a range of frequencies are cut or boosted in the form of a bell curve. In addition there are several forms of band adjusting filters, where an entire range of frequencies are adjusted uniformly. The icons typically used to indicate these filters accurately depict their functions.

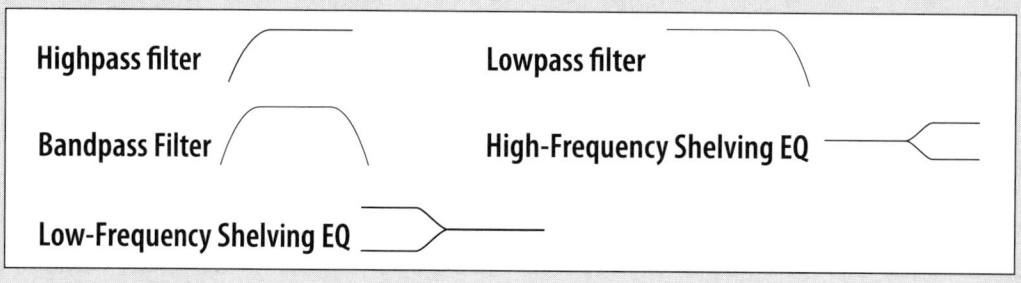

Highpass filter

Lowpass filter

Bandpass Filter

High-Frequency Shelving EQ

Low-Frequency Shelving EQ

Shelving EQ

A shelving EQ leaves all frequencies flat to a certain point, then turns all frequencies above or below that point down or up at a rate specified in dB per octave. As with high- and lowpass filters, shelving equalizers roll off the highs or lows, at a slope between 6 and 12 dB per octave; however, past the slope, all frequencies remain boosted or cut to the end of the frequency spectrum.

Shelving equalizers are a convenient way to add air (the high frequencies we can't necessarily hear as much as feel) to a mix. Simply sweep the cutoff frequency into the highs, above 12 kHz or so, and raise the shelf slightly. This is a common technique, especially with the advent of the ultra-quiet gear available today. In a previous era, these high frequencies spelled a slow death by tape noise.

Combined Equalizers

Software-based equalizers emulate each type of hardware EQ. Many contain identical controls and even emulate the look and feel of highly respected classic equipment. Additionally, several software-based equalizers offer all the features we've discussed, all in one equalizer. In fact, most offer multiple options at each band.

The Equalizer's Sound

Equalizing circuitry does affect the sound of the source. In fact, there are equalizers that sound good, and equalizers that don't sound so good. The quality, manufacturer, and design of any audio tool matters. Always listen to your music with and without the equalizer; be very discerning. It's better to avoid EQ than to use an EQ with lots of flexibility and a crummy sound.

Well-respected outboard gear gets that way because it works well and sounds good. Value reputation, yet always assess for yourself. Very inexpensive equalizers usually sound bad and are noisy—but not always. Very expensive equalizers usually sound great and are very clean and noise-free—but not always. It's up to you to listen and select the equipment that works for you.

Remember, even with the multitude of available equalizers, don't use EQ first to shape your sounds. First, get as close to the sound you want using mic choice and mic technique, then use EQ if it's necessary.

THE MONITOR SECTION

Some mixers have what is called the monitor section, which lets you listen to either the main outputs of the mixer or the audio recorder tracks as they're coming back to the mixer from the multitrack. The switch that selects where each control hears from usually has two positions—bus and recorder or tape, or sometimes input and recorder or tape.

This monitor section is used only for monitoring volumes and is totally separate from the recording level controls to the multitrack. Therefore, you can set exact levels to tape with the input faders, then turn their listening volume up or down in the monitor section.

Control Room Monitor Selector

The monitor selector is a very useful control center. It lets you listen to different buses, audio recorders, and playback devices in your setup simply by pressing the appropriate button on the board. This feature is usually located to the right of the channel faders.

Microphones & Mixers...by BILL GIBSON

If your mixdown recorder is normally connected to your mixer, if you have a CD player in your setup, and if you have one or more aux buses available, the monitor selector is a particularly valuable tool.

The monitor selector on most mixers lets you listen to different buses without affecting what's going on in the other buses, including the signal path to the multitrack. While recording a band's basic tracks, you can eavesdrop on the headphone bus, just to get an idea of what their mix sounds like, or you can listen to an effects bus to verify which instruments are being sent to a reverb or delay.

Each manufacturer designs what they feel is the best array of features at the price point that makes sense for them. Even the most modest mixer offers some of these monitoring options. However, professional consoles are designed to be fast and efficient, so most of these monitoring functions are included.

If you're using a Digital Audio Workstation you might not be using a mixer. There are a few manufacturers who currently provide audio control centers. These centers accept outputs from computers and playback devices; in addition, they can route various signals (and combinations of signals) to device inputs. They can also be used to select from among multiple monitoring systems. The PreSonus Central Station is an example of an excellent monitor controller that accepts multiple input formats and feeds multiple monitor systems. Some DAW controllers also provide many monitoring functions.

Stereo to Mono

The Stereo/Mono switch does just what it says. It lets you listen to your song in whatever stereo image you've created with the pan controls, or it can take your stereo mix and combine it all into one mono mix (meaning that exactly the same thing comes from both the left and right speakers).

This is very useful, especially if you expect that your song will be played on a mono system at any time. Mono is standard for AM radio, television, and live sound reinforcement. If you'll be playing your band's demo as break music at a performance, be absolutely sure that the demo sounds great even in mono. Audio Example 4-44 demonstrates a simple stereo mix.

· Audio Example 4-44

Simple Stereo Mix

Audio Example 4-45 uses the same mix, this time in mono.

· Audio Example 4-45

Stereo Mix in Mono

Notice the change in sound between Audio Examples 4-44 and 4-45. With some changes in panning and delay times, this mix can work well in stereo and mono.

Stereo Master

The stereo master control is the final level adjustment out of the mixer going to the mixdown recorder or power amp. The level adjustment to the mixdown recorder is very important. A good mix for a commercial-sounding song, in most styles, should be fairly constant in its level.

Adjust the stereo master fader for the optimum reading on the stereo output meter. This level should be set specifically for the mixdown recorder. The stereo master fader is not for volume adjustment—it is for final output level adjustment. Use the monitor level control for listening volume.

Monitor Control Centers

Control centers provide access to and from the various devices in your system. They are very useful and are typically well thought out and intelligently designed. The devices pictured here, the PreSonus Monitor Station and Central Station, are excellent examples of monitor control centers that integrate very well into a computer-based recording system.

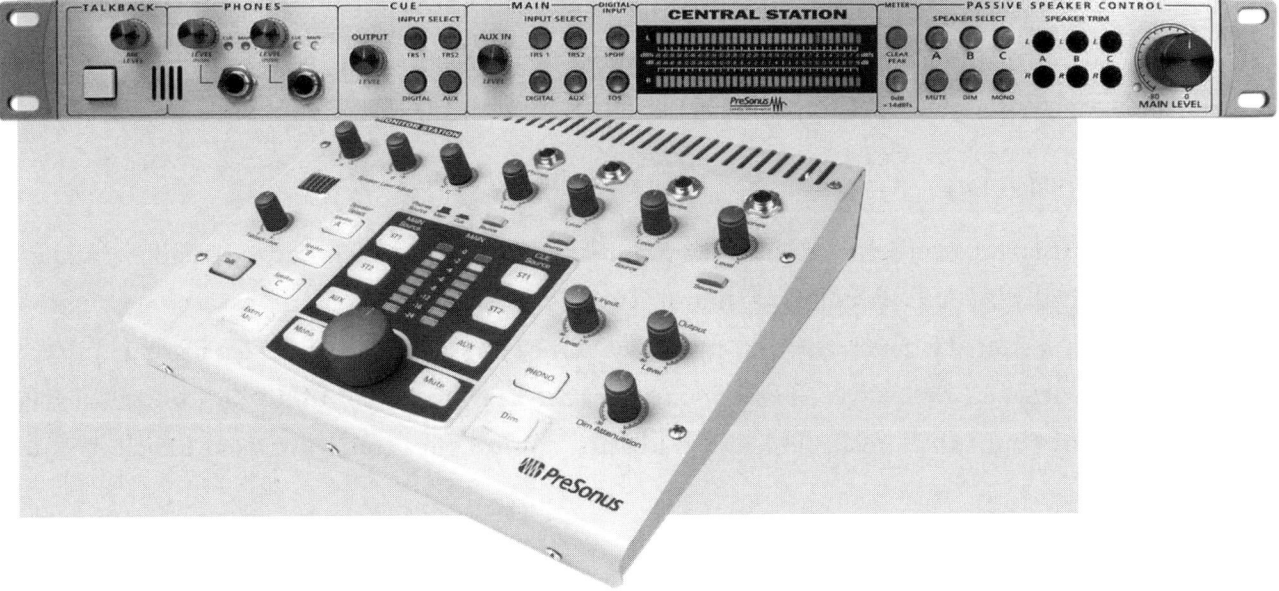

Microphones & Mixers... by BILL GIBSON

Monitor Level/Speaker Level

Use the monitor level control to set listening levels. In a recording setting, unlike live sound reinforcement, there's a difference between listening level and level to the multitrack or mixdown recorder. Set levels using the channel faders and the stereo master L/R fader. Adjust listening levels using the monitor level/speaker level control.

Talkback/Communications

The talkback button lets you talk to someone listening to the headphone bus. Communication with musicians through the headphones is essential to efficient recording. A small microphone is often mounted on the mixer for this purpose. Some mixers have a separate mic input for a handheld or stand-mounted talkback mic.

Typically, talkback can also be routed onto the multitrack or mixdown recorders to add a verbal reference like the song title, date, or artist. This verbal reference is called a slate. On vintage consoles, if a button has the word slate on it, there's typically a low frequency (around 40 Hz) that's sent to the recorder with your voice. In fast rewind or fast forward during the shuttle of analog tape, this low-frequency tone is heard as a beep, because the playback head picks up the magnetization as the tape speeds past the slate points. The slate beep is used to easily locate different songs on a reel.

Modern equipment really doesn't have much use for a slate tone, and most digital recording software offers excellent cataloging and locating features so that we don't necessarily need to print titles and the like to the multitrack mixdown recorder. There are, however, times when this ability is convenient and efficient, especially in the commercial studio. Hence, large format recording consoles provide the flexibility to route the talkback to multiple destinations.

Test Tones

Your mixer might have a section marked tones, test tones, osc, or oscillator, especially if you own a large format or vintage recording console. This section contains a frequency generator that produces different specific frequencies in their purest form—a sine wave. These frequencies are used to adjust input and output levels of your mixer, recorders, and outboard equipment. Tones are used for electronic

Chapter 4 .... SIGNAL PATH

169

calibration and level setting, whereas pink and white noise are used for acoustical adjustments

Consider the stereo master output from your mixer to the mixdown recorder. Raise the level of the reference tone (between 500 and 1,000 Hz) until the VU meter reads 0 VU on the stereo output of the mixer. Do this with the stereo master output faders set at the point where your mix level is correct. Adjust the level to the meters with the tone's output control.

Fine-tune the left/right output balance. If one side reads slightly higher than the other, from the same 1-kHz tone, balance the two sides. Many mixers have separate level controls for left and right stereo outs. Proper adjustment of the left/right balance ensures the best accuracy in panning and stereo imaging.

Modern digital consoles and digital recorders are very stable in their levels and typically hassle free as far as setting machine-to-machine recording levels. However, high-quality interfaces, such as Pro Tools HD I/Os require setup and calibration to function at the peak performance.

INTEGRATING ANALOG AND DIGITAL RECORDERS

It is likely that you will eventually want to integrate digital and analog devices. There are a few key considerations in each area that are non-issues in the other. Spend some time learning about both digital and analog recording tools. They're both very valid in the world of professional audio production.

Analog Reference Tones

One thousand Hz is the most common reference tone. A reference tone is an accurate representation of the average recording level. Therefore, if your mix level is correct and peaks at 0 VU or +1 or so, then this 1-kHz tone at 0 VU is an accurate gauge for setting levels for duplication of this particular song.

Patch the output of the mixer to the mixdown recorder and adjust the input level of the mixdown recorder to read 0 VU while the mixer is showing 0 VU from the 1-kHz tone. We can now be sure that the level on the board matches the level on the mixdown machine.

Once these levels match, go ahead and record some of the 0 VU to your mixdown recorder so that when you make copies, you can use this as a reference tone to set the input levels of the duplicating machine.

170

Microphones & Mixers .... by BILL GIBSON

If you're printing your entire mix to analog tape, there are certain procedures you should follow:

  • Be sure your mixdown recorder is aligned and calibrated to work optimally with the specific brand and formulation of tape you're using. Refer to your owner's manuals for specifics to perform this setup or, better yet, hire a professional tech to get the analog recorder set up and matched to your mixer levels.

  • Patch the mixdown master output from your mixer to the line inputs of your mixdown recorder.

  • Set the mixdown machine to record ready and, if available, select monitoring of the input or source.

  • Set 0 VU as I specified above so that the mixer output and the mixdown recorder input read 0 VU.

  • If your mixer has a tone generator or frequency oscillator, turn it on, select a frequency between 500 Hz and 1,000 Hz, and raise the level on the mixdown master VUs on the mixer until they read 0 VU.

  • At the beginning of the first reel of mixes, record the 0 VU tone. This is called a reference tone. If your master will be duplicated by a professional duplication facility, these tones let them adjust the level of their equipment to match yours. Following this procedure should result in a better, cleaner, and more accurate copy.

Next, record a series of tones that represent high, mid, and low frequencies. The standard frequencies to record at the beginning of a master tape are 100 Hz, 1 kHz, and 10 kHz. Giving the duplicator these references helps them compensate for any inherent problems in your equipment.

PATCH BAYS

A patch bay is nothing more than a panel with jacks in the front and jacks on the back. Jack #1 on the front is connected to Jack #1 on the back, #2 on the front to #2 on the back, and so on.

If all available ins and outs for all of your equipment are patched into the back of a patch bay and the corresponding points in the front of the patch bay are clearly labeled, you'll never need to search laboriously behind equipment again just to connect two pieces of gear together. All patching can be done with short, easy-to-patch cables on the front of the patch bay.

Chapter 4 .... SIGNAL PATH

171

Patch Bays

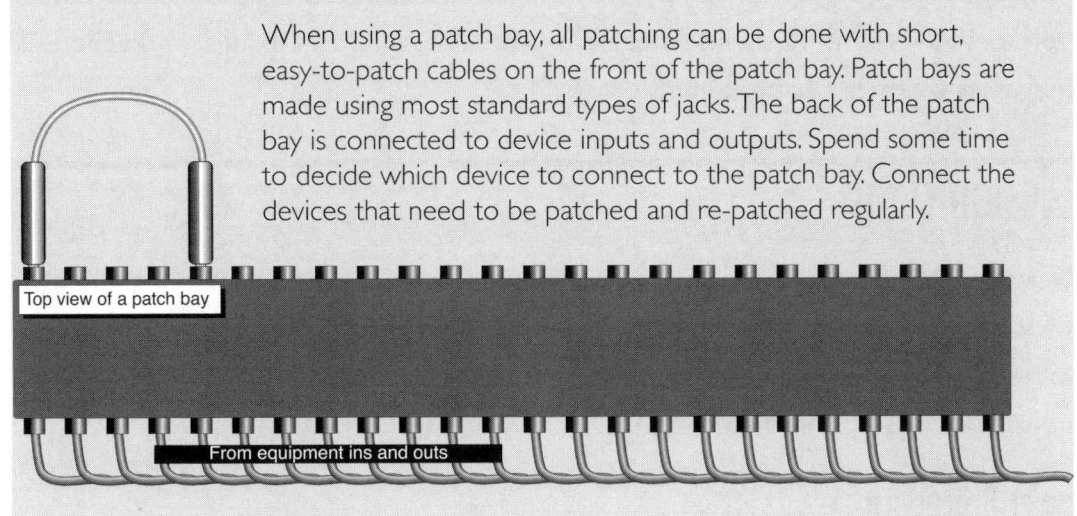

When using a patch bay, all patching can be done with short, easy-to-patch cables on the front of the patch bay. Patch bays are made using most standard types of jacks. The back of the patch bay is connected to device inputs and outputs. Spend some time to decide which device to connect to the patch bay. Connect the devices that need to be patched and re-patched regularly.

Top view of a patch bay

From equipment ins and outs

Patch bays are used for line-level patches like channel ins and outs, tape recorder line ins and outs, sound module outputs, and any signal processor ins and outs. Don't use the patch bay for powered outputs, like the speaker outputs of your power amplifiers.

The concept of easy and efficient patching becomes obvious when it's explained, and once you've made the move to include a patch bay in your setup, you'll be able to accomplish more, faster and more efficiently.

Most commercial recording facilities utilize patch bays extensively, simply because they are fast and efficient. When there are a few different types of projects moving through a studio each day, there's great value in quick and easy setups—this is the strength of the patch bay.

There is, however, a disadvantage to incorporating a patch bay into your system at home—it can degrade the sound quality. Patch bays are likely to rob something from the sound you're recording, especially if they're not impeccably cleaned and maintained or a regular basis. Anytime you plug a cable into almost anything, you risk some signal degradation.

Microphones & Mixers..by BILL GIBSON

Even when I'm tracking or mixing in a world-class recording facility, I eliminate any patch points that aren't absolutely necessary. I don't mind running a cable from one room to another, or under a door, or down the hall, because I've heard the difference in sound quality when I bypass unnecessary patches.

At home, you need to decide whether incorporating a patch bay is appropriate. If it makes your setup more efficient, in a way that enhances the creative freedom of your recording environment, that's the bottom line. I'm all for great sound, but what we do is all about great music.

SESSION SETUP

Use this procedure as a starting point for your sessions. Start each session with your studio clean and all equipment adjusted to a predetermined typical level. Starting clean prevents problems resulting from unknown buttons being pushed in unknown places on the mixer.

Basic Procedure

*   Move all channel faders to 0.
*   Set input gain (preamp) and attenuator to lowest level.
*   If you have only a mic/line switch, set it to line.
*   Pan all channels to center.
*   Set all EQ to flat (no boost or cut). If there is an EQ in/out switch, set it to out.
*   Turn any auxiliary sends, effects sends, or reverb sends all the way down or off.
*   Set VU meters to allow monitoring of the final stereo output to the mixdown machine. If available, set other VU meters to monitor levels of aux buses to effects.
*   Make sure there are no track assignments selected.
*   Be sure there are no solo buttons selected.
*   Be sure there are no mutes selected.

Chapter 4 ....SIGNAL PATH

173

NEW MIXER FUNCTIONS

Digital audio workstations, audio recording software, FireWire and USB interfaces, and control surfaces have radically changed the way we record audio. Though we'll study new technology in this series, the fundamental concepts in this chapter provide the necessary knowledge and insight that you'll need to build a powerful understanding of general audio recording.

Being well versed in the recording world is the only way you'll be able to communicate with other enlightened musicians and engineers, and that constant communication can inspire your individual growth and possibly open doors into the business world of music and recording.

The mixer can be your most flexible means of achieving the musical sounds that you want. Go to your own setup and find what kind of controls you have. Review this material thoroughly and apply each point, deliberately, to your own setup.

A thorough understanding of the information in this chapter is necessary as a foundation for upcoming chapters. We'll build on this foundation in a methodical, easy-to-follow way. Each chapter is structured, using combined media, to closely resemble a private lesson.

If you do the assignments and study the DVD-ROM, text, and illustrations, you'll see a marked difference in your recording skills and end results.

174

Microphones & Mixers .... by BILL GIBSON

CHAPTER TEST

1. For speed and efficiency in any recording situation, it's essential that you're completely familiar with the _____ involved in your setup.
   a. impedance
   b. signal paths
   c. variables
   d. control points

2. A block diagram:
   a. is a map of the entire recording system
   b. contains the fundamental elements and building blocks of the the recording source
   c. is a line drawing consisting of the signal path layout and design for a specific piece of equipment
   d. All of the above

3. To adjust the mixer channel input gain trim:
   a. turn up the trim until the peak indicator blinks during a loud passage, and then turn the trim back a little so the peak indicator only blinks a couple times during a recording pass
   b. set the input fader at unity position and adjust the trim for a proper recording level
   c. press the input channel SOLO button so that the input level registers on the level meter and adjust the trim for a proper input level
   d. None of the above
   e. All of the above

4. Sometimes, the signal from the microphone is too hot, no matter where the trim is set. In this case use the channel input attenuator or the built-in attenuator on the microphone to compensate.
   a. True
   b. False

5. The signal-to-noise ratio specifies the distance, in decibels, from the noise floor to the loudest sound in an audio signal.
   a. True
   b. False

6. When the input trim is set too low:
   a. excessive noise combines with the intended signal
   b. the impedance mismatch results in an unacceptable level
   c. the source becomes distorted because of the level mismatch
   d. All of the above

7. Whenever two or more microphones pick up the same waveform, there is a likelihood that the resulting sound will be superior because of the increased response characteristic.
   a. True
   b. False

8. The channel phase reversal switch turns the signal around so that it bounces back to the source.
   a. True
   b. False

9. Frequently, several channel inserts are combined through a summing bus to provide multiple sends to reverberation and delay effects.
   a. True
   b. False

10. To utilize a single-jack insert, you must use a special Y cable with a TRS connector for the mixer insert and two TS connectors for the input and output of the device.
    a. True
    b. False

11. The best way to route signals to effects devices is through the use of:
    a. an insert
    b. a specially constructed Y cable
    c. a PRE bus
    d. an aux bus
    e. All of the above

Chapter 4 .... SIGNAL PATH

175

12. Pressing the PRE button on a specific aux send:
    a.   accesses the signal before the input preamp
    b.   accesses the signal before the channel insert
    c.   accesses the signal just before the channel mix level fader
    d.   accesses the signal just before it is routed to the mixdown recorder

13. Pressing the POST button on a specific aux send:
    a.   accesses the signal just after the input preamp
    b.   accesses the signal just after the channel insert
    c.   accesses the signal just after the channel mix level fader
    d.   accesses the signal just after it is routed to the mixdown recorder

14. Aux buses are most appropriate for use with _____. Inserts are most appropriate for use with _____.
    a.   instruments, vocals
    b.   delay effects, dynamics processors
    c.   reverberation, compressors
    d.   mixdown, tracking
    e.   b and c

15. When a specific frequency is boosted or cut on an equalizer, that frequency is the center point of a curve being boosted or cut.
    a.   True
    b.   False

16. Narrow Q settings are typically best suited for _____, whereas wide Q adjustments are best suited to _____.
    a.   tracking, mixdown
    b.   repairing problems, tone control
    c.   vocals, instruments
    d.   a and b

17. A parametric equalizer is the most flexible of the equalizer types because it contains:
    a.   a bandwidth control
    b.   adjustment of the Q
    c.   the ability to select various frequencies
    d.   a and b
    e.   All o f the above

18. A bandpass filter:
    a.   is any filter that affects a frequency band
    b.   lets a portion of the frequency spectrum above and below specified points pass through unaffected
    c.   cuts the highs above a certain point in the frequency spectrum
    d.   cuts the lows below a certain point in the frequency spectrum

19. A highpass filter:
    a.   is any filter that affects a frequency band
    b.   lets a portion of the frequency spectrum above and below specified points pass through unaffected
    c.   cuts the highs above a certain point in the frequency spectrum
    d.   cuts the lows below a certain point in the frequency spectrum

20. A patch bay:
    a.   is a device that provides convenient access to common device inputs and outputs
    b.   makes patching between devices very simple
    c.   can degrade the audio signal and should be avoided if possible to ensure maximum signal integrity
    d.   All of the above

Test answers are on page 293

Microphones

There's much more to mic choice than finding a trusted manufacturer that you can stick with. There's much more to mic placement than simply setting the mic close to the sound source. The difference between mediocre audio recordings and exemplary audio recordings is quite often defined by the choice and placement of microphones.

The microphone is our most fundamental tool. You can have $100,000 worth of esoteric, vintage, high-tech gear in your signal path, but if the microphone doesn't capture the sonic essence that's perfect for the recording, it's all a waste. Each microphone offers a sonic personality and offers the potential to be much more than just an archival tool. For instance, if you and a buddy test 10 mics on Joe to see how they'll work for his new song, nine of the microphones might evoke agreement that, yeah, that sounds like Joe. However, chances are that one of the 10 mics might get the response, "Wow! Joe sounds great on this mic!"

Once you find the microphone that sounds great for whatever sound source you're recording, it's time to compare other options in the signal path, such as preamplifiers, compressors, or equalizers—the fewer additions to the signal path, between the mic and the recorder, the better. Musically, you need to do what you need to do. As long as it feeds the passion and emotion of the music, it's alright to include a hundred processors in the signal path; however, if you want to capture the true essence of the

178

Microphones & Mixers.. by BILL GIBSON

original sound at the source, find the perfect mic and the perfect preamp, patch directly into the recorder, and record.

If you need to use compression, make it a conscious choice, and be sure the compressor is enhancing the musical impact. In the modern era, noise isn't really an issue, so you can always add compression and other processing during mixdown. Save as many musical decisions as possible for mixdown.

························· **Video Example 5-1**

Several Different Microphones on Acoustic Guitar

························· **Audio Example 5-1**

Several Different Microphones on Acoustic Guitar

As we cover techniques for recording different instruments, we'll consistently need to consider microphone choice and technique. The MIDI era led us away from the art of acoustic recording, but as time has proceeded, acoustic recordings of drums, guitars, strings, brass, percussion, and sound effects have returned. There is a kind of life to an acoustic recording that can only truly exist through recording real instruments played by real people in an acoustical environment.

We can increase the life in our MIDI sounds by running them into an amplification system, then miking that sound. We might or might not need to include the direct sound of the MIDI sound module.

Using a mic to capture sound is not as simple as just selecting the best mic. There are two other critically important factors involved in capturing sound:

+ Where we place the mic in relation to the sound source
+ The acoustical environment in which we choose to record the sound source

As you'll see in the audio examples in this course, the sound of the acoustical environment plays a very important role in the overall sound quality.

Selecting a microphone involves more than a simple random search for "the sound." An understanding of mics and how they work, along with the ability to read and understand their specifications, allows us to make educated predictions regarding which mics to consider. You should be able to listen to a source, and then make an intelligent decision about which mics to audition.

Technically, we must consider a set of factors when choosing a microphone: directional characteristic, operating principle, response characteristic, and output characteristic. In addition, the real-world considerations are always cost, durability, and appearance. In today's market, there are several very inexpensive mics available that look great. Some of them sound okay, but if you need some impressive looking mics for your clients' sake, consider them like a decoration. However, if you intend to make great music and are serious about your craft, always be in search of great-sounding microphones.

DIRECTIONAL CHARACTERISTIC

Any time you mic a source, you must be aware of which way to aim the mic, and whether that particular mic is sensitive only to the intended source. Understanding the microphone polar response pattern, also called the pickup pattern, is fundamental to capturing the essence of the sound you intentionally want. There are three basic polar patterns we consider when comparing and choosing microphones: omnidirectional, bidirectional, and unidirectional.

We typically refer to the overall pickup in our discussions; however, each polar response will vary dramatically when frequency ranges are compared. In addition, there are variations of the directional characteristics. Because of the way most directional characteristics are created, high frequencies are the most directional and low frequencies are the least directional.

The final polar response pattern is not usually a result of the inherent operational principle of the microphone capsule. More often, it's either the result of

Polar Patterns

Polar graphs are seen in two dimensions, but they imply three dimensions. All patterns should be visualized in a spherical three-dimensional plane.

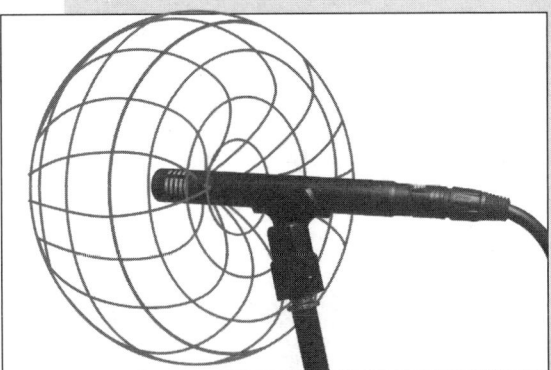

Microphones & Mixers .... by BILL GIBSON

the physical housing design, the electronic balance of multiple diaphragms, or a combination of both the physical and electronic design characteristics.

Polar Response Graph

The polar response graph provides a visual image of the microphone's sensitivity to sound coming from different directions. This circular graph indicates sensitivity in a 360-degree circular scope and is interpreted as a three-dimensional image, even though it's drawn two-dimensional for the sake of simplicity.

On-Axis

Zero degrees, on the graph, represents the front of the microphone (the portion designed to pick up the sound). This position is referred to as on-axis.

Off-Axis

Any position on the polar response graph that isn't on-axis is called off-axis and is quantified in degrees. 180 degrees, on the graph, represents the back of the mic (the part directly in back of the portion designed to pick up the sound); it's referred to as 180 degrees off-axis.

Polar Response Graph

The polar response graph plots the spatial sensitivity as it relates to the position of the sound source in relation to the microphone capsule. These graphs are considered symmetrical in relation to the plotted sensitivity and, in addition, should be considered three-dimensionally spherical.

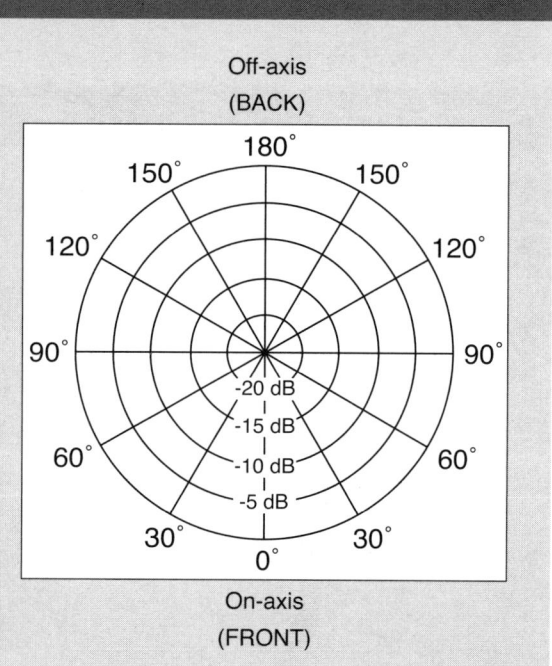

Polar Shapes

The two most basic polar response patterns are omnidirectional (doesn't discriminate against sounds from any direction) and cardioid (discriminates against sounds that are 180 degrees off-axis). The other two polar shapes in this illustration are bidirectional (an omnidirectional pattern on each side of the mic) and hypercardioid (a bidirectional pattern with a large half and a small half).

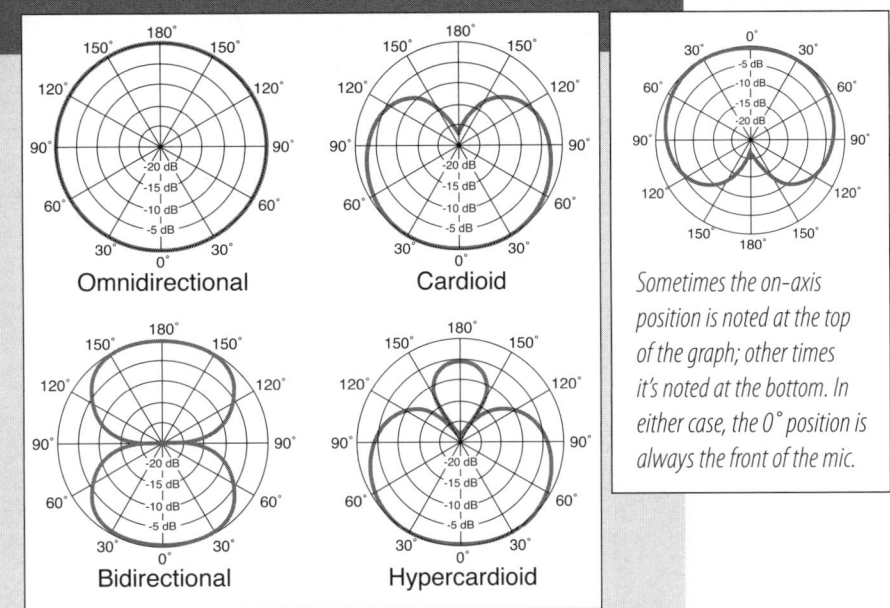

Omnidirectional

Cardioid

Bidirectional

Hypercardioid

Sometimes the on-axis position is noted at the top of the graph; other times it's noted at the bottom. In either case, the 0° position is always the front of the mic.

A microphone, which demonstrates a decrease in sensitivity at a certain point on the polar graph, is said to discriminate from sound at that point. We indicate that decrease in sensitivity by denoting its degree marker. A microphone with reduced sensitivity at 180 degrees on the graph is said to exhibit 180-degree off-axis discrimination.

Additionally, for the sake of indicating discrimination the graph is considered symmetrical. Indicating that a microphone exhibits 150-degree off-axis discrimination indicates that the point of least sensitivity is at 150 degrees, and since the graph is symmetric, also at 210 degrees. This also indicates that there is sensitivity, to some degree, in the region centered on 180 degrees off-axis.

Each microphone uses a design that exhibits a unique polar response throughout the frequency spectrum.

Because of the symmetrical aspect of the polar graph, sometimes they're indicated 0–180 degrees left and right instead of 360 degrees around. This is actually a little simpler system, since off-axis discrimination is always referenced as 180 degrees or less, with the symmetry across the 0 degree axis implied.

Sensitivity Scale

The polar response graph calibrates spherical sensitivity through a series of concentric circles. Each concentric circle is consecutively smaller. The outer circle

182

Microphones & Mixers... by BILL GIBSON

Multiple Frequencies on the Polar Graph

The polar graph often displays the directional characteristic for multiple frequencies. To accomplish this in the least cluttered manner, all patterns are assumed to be symmetrical across the Y axis. In addition, to help clarify the results, various line styles are incorporated on each frequency. Sometimes the polar graph is split, like the graph on the left, to highlight the variations in frequency response; other times the graph is whole, like the graph on the right, with the pattern variations simply changing between left and right.

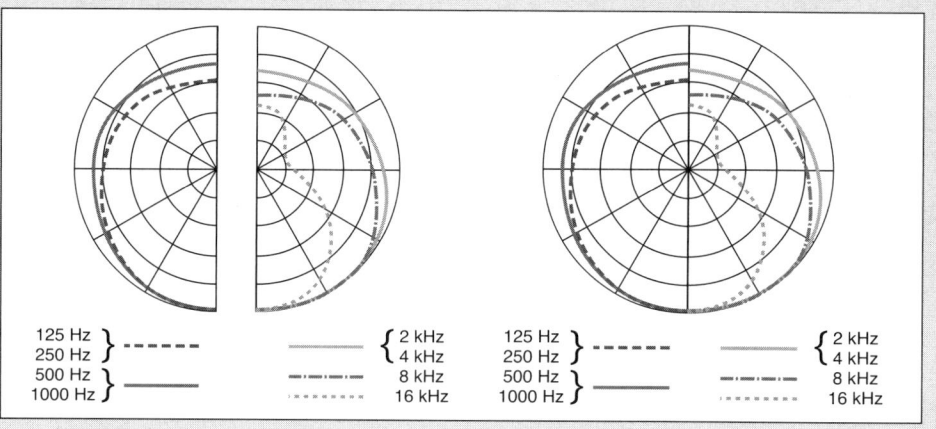

125 Hz				2 kHz
250 Hz	} -------		———	{ 4 kHz
500 Hz	}		—·—·—·	8 kHz
1000 Hz	}		·········	16 kHz

125 Hz				2 kHz
250 Hz	} -------		———	{ 4 kHz
500 Hz	}		—·—·—·	8 kHz
1000 Hz	}		·········	16 kHz

indicates full sensitivity (0-dB decrease in sensitivity), and there are typically four or five consecutively smaller circles between the outer circle and the center point, each typically indicating a decrease in sensitivity of 5 dB. Plotting the decreases of sensitivity around the polar graph is what creates the polar pattern.

Some polar graphs include an additional outer circle, at +5 dB, for the rare instance that certain frequencies sum, creating a hypersensitivity, which exceeds the normal full sensitivity.

Normally, the pattern on the graph, which is visually dominant and uses a solid bold line, is the average overall pattern and is typically measured from a 1-kHz sine wave. Many electronic measurements consider a 1-kHz sine wave as the reference, or average, signal across the audible spectrum.

Omnidirectional

An omnidirectional mic, sometimes referred to simply as an *omni* mic, is equally sensitive to sounds in all directions. Relative to mics with other directional characteristics, omnidirectional mics are the least negatively affected by the proximity effect. They don't reject sound from any direction and they typically provide the fullest and most natural sound when used in a distant-miking application.

Omnidirectional

Microphones with an omnidirectional pickup pattern pick up sound equally from all directions and don't reject sound from any direction.

Keep in mind that directional characteristics are three dimensional, so an actual omnidirectional pickup pattern is in the shape of a globe.

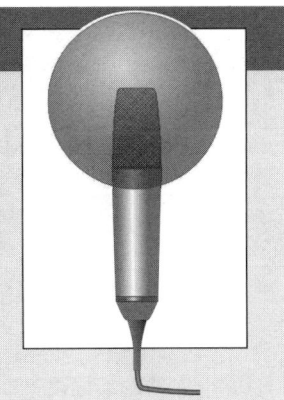

An omnidirectional microphone is a very good choice for capturing room ambience, recording groups of instruments that can gather around one mic, close-miking a vocalist, and capturing a vocal performance while still letting the acoustics of the room interact with the sound of the voice.

Omnidirectional microphones are inappropriate in a live setting because they produce feedback more quickly than any other pickup pattern.

Bidirectional

Bidirectional microphones hear equally from the sides, but they don't hear from the edges. Bidirectional microphones are an excellent choice for recording two sound sources to one track with the most intimacy and the least adverse phase interaction and room sound. Position the mic between the sound sources for the best blend. Just beware that, once you've recorded the track, there's not much you can do in the mix to fix a bad balance or blend.

Bidirectional

Bidirectional microphones hear equally well from both sides, but they don't pick up sound from the edge.

Bidirectional mics work very well for recording two voices or instruments to one track. This is also called a figure-eight pattern.

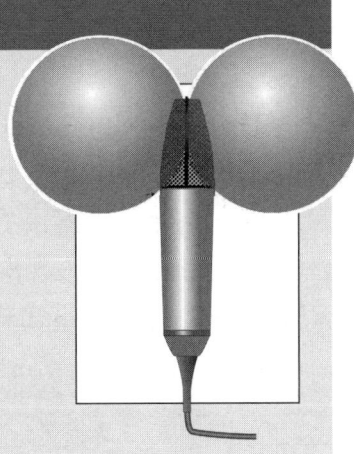

184

Microphones & Mixers... by BILL GIBSON

Cardioid Pickup Pattern

A microphone with a cardioid pickup pattern hears sound best from the front and actively rejects sounds from behind. With its heart-shaped pickup pattern, you can point the mic toward the sound you want to record and away from the sound you don't want to record.

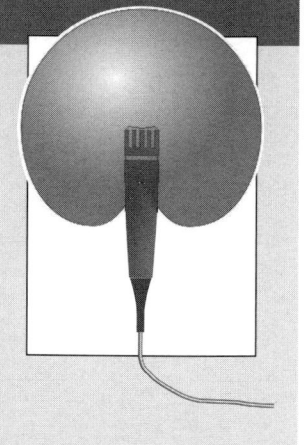

Unidirectional

Most microphones demonstrate a unidirectional characteristic, often called a directional or cardioid pickup pattern. The pickup pattern is visually represented by a heart-shape—rounded in front and dimpled in the back. The unidirectional mic is most sensitive (hears the best) at the part of the mic into which you sing; it's least sensitive (hears the worst) at the side opposite the part into which you sing.

The advantage to using a microphone with a cardioid pickup pattern lies in the ability to isolate sounds. You can point the mic at one instrument while you're pointing it away from another instrument. The disadvantage to a cardioid pickup pattern is that it will typically only give you a full sound from a close proximity to the sound source. Once you're a foot or two away from the sound source, a cardioid pickup pattern produces a very thin-sounding rendition of the sound you're miking.

In a live sound setting, directional mics are almost always best because they produce far less feedback than mics with omnidirectional or bidirectional pickup patterns.

There are five unidirectional pickup patterns for consideration in normal use: cardioid, supercardioid, hypercardioid, ultracardioid, and subcardioid. Here is a comparison of their fundamental polar response patterns. Each microphone is unique in design and may exhibit its own rendition of these response patterns. Also, keep in mind that, throughout the frequency spectrum, results may vary.

Chapter 5 .... MICROPHONES

185

Cardioid

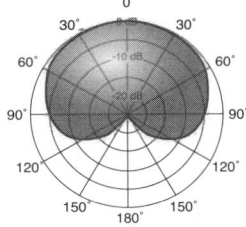

The cardioid pickup pattern demonstrates full response at the front of the microphone and a decrease in sensitivity of up to 25 or 30 dB at 180 degrees off-axis.

In relation to a cardioid pickup pattern, the supercardioid and hypercardioid pickup patterns each become progressively narrower on the sides, with an increased area of off-axis sensitivity.

Supercardioid

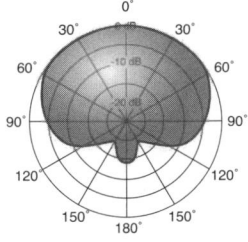

A microphone with a supercardioid polar response is more directional at the front than a microphone exhibiting a cardioid pattern, with a decreased sensitivity on the sides and an area of sensitivity about 170 degrees off-axis.

Hypercardioid

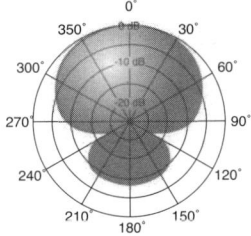

A microphone with a hypercardioid polar response exhibits a high degree of directionality at the front, with a decrease of about 12 dB on the sides and an area of least sensitivity at about 110 degrees off-axis.

Ultracardioid

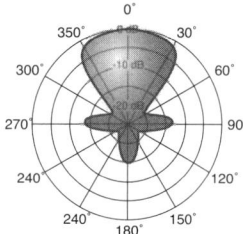

A microphone with an ultracardioid polar response is very focused and directional in front with a small area of sensitivity at 90 degrees and 180 degrees.

Subcardioid

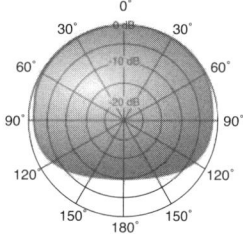

The subcardioid polar response is wider and extends further than the cardioid pattern, approaching the non-directionality of an omnidirectional microphone.

OPERATING PRINCIPLE

Although there are hundreds of different microphones available from a lot of manufacturers, they essentially all fit into three basic categories: condenser, moving-coil, and ribbon. Condenser and moving-coil mics are the most common of these three, although they may all be used in recording, as well as live, situations.

There are other types of microphones with operating principles that differ from what we will cover in this course, and each type of microphone has its own individual personality. Mic types other than condenser, moving-coil, and ribbon are usually selected for a special effect in a situation in which the music needs a unique sound that enhances the emotional impact of the song.

Transducer Types

A transducer is any device that transforms one type of energy into another type of energy. For instance, a speaker converts electrical energy into acoustic energy. The amplifier sends an electronic signal (a continuously varying flow of electrons) to the speaker, which responds to the electronic signal by moving air, which is, at that point, acoustical energy. Your ear is another example of a transducer because it converts the acoustic energy into electrical energy, which is then sent to the audio perception portion of your brain.

Whether in a recording or live sound application, the microphone is the first transducer, past the source, in our signal path. The microphone converts acoustic energy into electrical energy. There are other transducers that might have been involved prior to the actual sound source. For our purposes in audio, we'll consider the microphone as the first transducer within our control.

Of the three types of microphones we'll study, there are only two types of transducers: magnetic induction and variable capacitance. The transducer is the actual microphone capsule—the point at which the acoustic energy from the sound source reaches the mic and begins the flow of electrons. The microphone might contain amplifying circuitry, which is insignificant in our understanding of transducer functionality.

Magnetic Induction Transducers

A magnetic induction transducer utilizes a process in which metal, which has magnetic properties (in other words, it can be magnetized), is stimulated into motion around, or is attached to, a magnet. Of the three mic types that we'll study, two use magnetic transducers: moving-coil and ribbon.

Mics that use magnetic induction, whether moving-coil or ribbon, are dynamic microphones. Often, moving-coil mics are generically referred to as dynamic microphones, and ribbon microphones are differentiated as...ribbon microphones. It is more accurate to differentiate them as moving-coil and ribbon mics.

Variable Capacitance Transducers

Variable capacitance transducers operate on an electrostatic, rather than a magnetic, principle. A variable capacitance microphone capsule utilizes a fixed, solid conductive plate adjacent to a flexible piece of plastic that's been coated with a conductive alloy, often containing gold.

A capacitor is a device that stores an electrical charge. When the movable plate is electrically charged, an electrostatic charge is stored between the two

surfaces. As sound waves vibrate the alloy-coated plastic diaphragm, the area of stored electrical charge emits a continuously varying flow of electrons that accurately portray the waveform. Variable capacitance transducers require power to charge the plates and to power an amplifying circuit within the microphone.

A microphone that uses a variable capacitance transducer is sometimes called a capacitance microphone, although more often it's referred to as a condenser microphone. The old-school name for a capacitor was condenser—same device, different name.

OPERATING PRINCIPLE OF THE MOVING-COIL MIC

A moving-coil microphone, which is frequently referred to as a dynamic microphone, operates on a magnetic principle. When an object that can be magnetized is moved around a magnet, there is a change in the energy within the magnet. There is also a continual variation in the magnetic status of the object moving in relation to the magnet. The moving-coil microphone uses this fact to transfer the

Moving-Coil Microphones

Copper wire is wrapped into a cylinder. This cylinder is then suspended around a magnet. The copper coil moves up and down in response to pressure changes caused by sound waves.

The crest of the audio wave moves the coil down, causing a change in the status of the magnet. The trough of the audio wave moves the coil up, again causing a change in magnetism.

As the coil moves up and down, the magnet receives a continually varying magnetic image. The continually varying magnetism will ideally mirror the changing air pressure from the sound wave. This continually varying magnetism is the origin of the signal that arrives at the mixer's mic input.

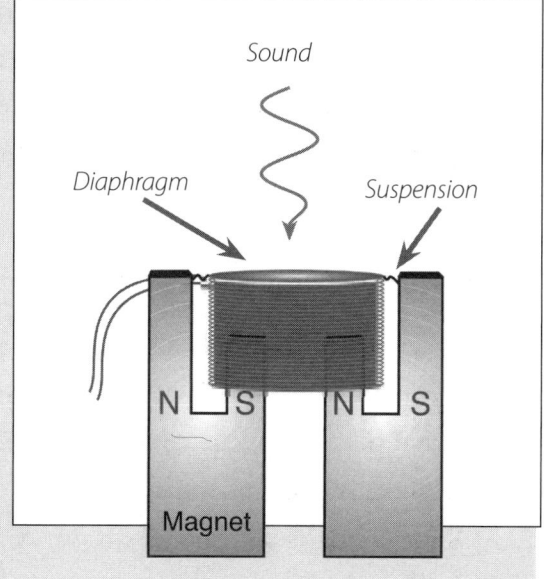

changing air pressure, produced by an audio waveform, into a continually varying flow of electrons that can be received by the mic preamp.

In a moving-coil mic, a coil of thin copper wire is suspended over a fixed magnet, enabling the coil to move up and down around the magnet. A thin Mylar plastic diaphragm closes the top of the coil and serves to receive the audio waves. As the crests and troughs of the continually varying audio waveform reach the diaphragm, the coil is forced to move around the magnet. The movement of the copper coil around the magnet is what causes the changing flow of electrons that represents the sound wave.

OPERATING PRINCIPLE OF THE RIBBON MIC

A ribbon microphone operates on a magnetic principle like the moving-coil. A metallic ribbon is suspended between two poles of a magnet. As the sound wave vibrates the thin ribbon, the magnetic flow changes in response, causing a continually varying flow of electrons. As the ribbon moves between the poles of the magnet, it varies the status of the magnet's north and south magnetism in direct

Ribbon Microphones

A thin metal ribbon suspended between two poles of a magnet vibrates in response to each crest and trough of a sound wave. As the ribbon moves in the magnetic field, it continually varies in its magnetism. These changes of magnetism are the origin of the signal that is sent to the mic input of your mixer.

The signal produced by the ribbon is typically weaker than the signal produced by the moving-coil. In practical terms, that means you'll usually need more preamplification at the mic input to achieve a satisfactory line-level signal.

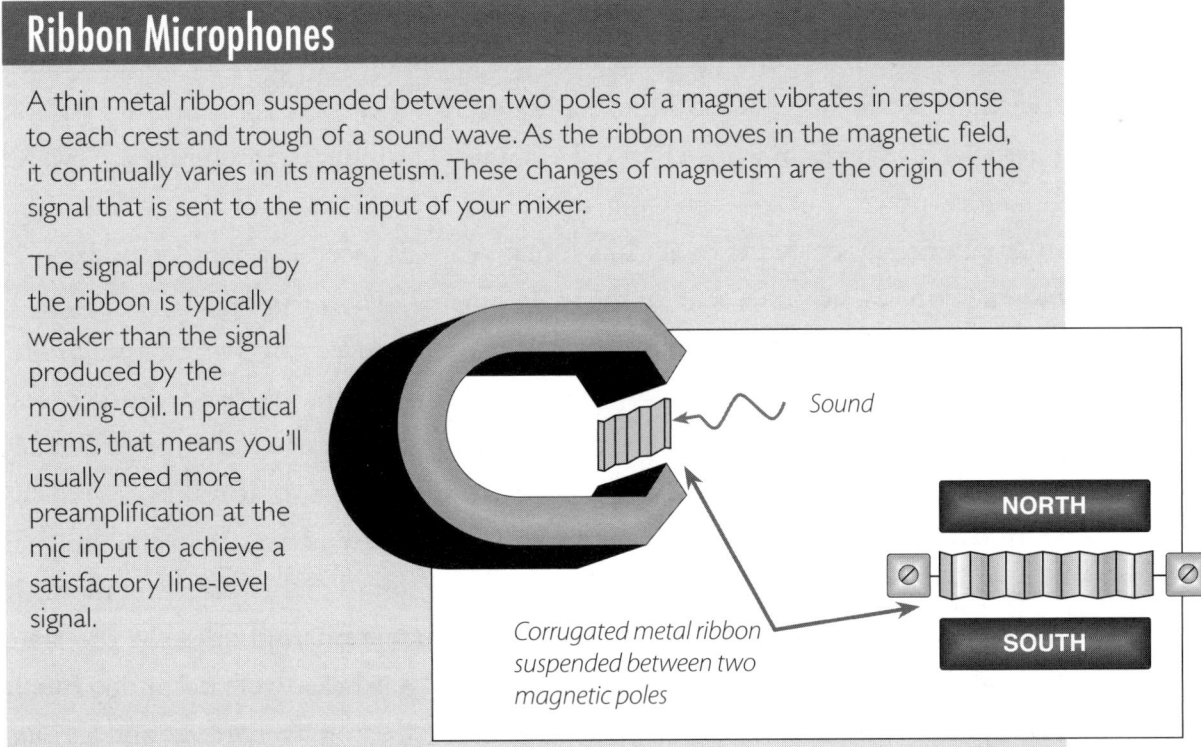

Sound

NORTH

SOUTH

Corrugated metal ribbon suspended between two magnetic poles

Microphones & Mixers.. by BILL GIBSON

proportion to the changes in amplitude produced by the sound wave. This continually varying flow of electrons is the origin of the signal that reaches the microphone input of your mixer.

Historically, ribbon microphones have been very fragile; the ribbons needed to be a certain length to generate enough signal strength, and they needed to be thin enough to respond accurately to sonic nuance. Therefore, vintage ribbon mics, such as the RCA 77 DX, constantly need maintenance, although they sound great when they're functioning up to specifications. For several years, the recording industry mostly ignored ribbon microphones. Since the classic RCA 77 DX and 44 BX were a standard in the early and mid-1900s, there have been a few popular ribbon microphones, including the Beyer M500, Beyer M160, and the Shure SM33 (popular for sitting on Johnny Carson's *Tonight Show* desk for many years). All of these microphones exhibit a unidirectional characteristic, created by the physical housing around the mic capsule, rather than the inherent bidirectional characteristic of the ribbon microphone capsule.

Modern ribbon microphones are capable of using smaller, stronger magnets, which enables the use of shorter ribbons. This has resulted in the production of more durable ribbon mics, although they still are the most fragile of the three types we use.

Because ribbon microphones utilize a permanent magnet in their capsule, they don't require a power source to operate, although there are some new ribbon mics that use phantom power to drive internal amplification circuitry. Beware! Phantom power can be damaging to ribbon mics other than these. A vintage microphone ribbon tends to act like a fuse when it receives phantom power—it blows. (Pop, kaput, gone, send in for replacement—get my point?)

In 1998, David Royer and Royer Labs introduced the R-121 and SF-12 ribbon microphones—this was an extremely timely offering to the burgeoning digital recording industry and, in many ways, a game changer. The Royer R-121, a true bidirectional mono mic, and the SF-12, a stereo mic that utilizes two bidirectional capsules aimed 90 degrees apart, instantly found a place in the modern recording world. A ribbon mic capsule provides a sound that is warmer and much less strident than condenser or moving-coil dynamic microphones. At the time they were released, digital recording software and hardware were becoming much more powerful, widely accepted, and cost effective for home recordists and serious professionals who wanted to work from home. However, at the time there was a realization that digital technology, although clean and noise-free, had lost a lot of the warmth and smoothness captured by classic analog technology. Warmth and smoothness are prime sonic characteristics of ribbon microphones! Royer has developed many iterations of the ribbon microphone and their mics have become widely accepted as a necessary tool in the modern recording professional's bag of tricks. They have also become popular in the live sound reinforcement industry because they deliver a smooth, warm tone that is less grinding and abrasive, especially when miking aggressive sounds such as electric guitars, drums, stringed instruments, and brass.

OPERATING PRINCIPLE OF THE CONDENSER MIC

A condenser mic operates on a fairly simple premise, although it is physically based on a different principle. Whereas the moving-coil and ribbon microphones operate on a magnetic inductance principle, the condenser mic is based on a variable capacitance principle.

A voltage (positive or negative) is applied to a metal-coated piece of plastic. The plastic is a little like the plastic wrap you keep on your leftover food. The metallic coating is thin enough to vibrate in response to sound waves; in fact, the technique used to apply the coating to the membrane is called sputtering because the alloy is so lightly applied. Its function is to provide conductivity for the electrical charge while not inhibiting the flexibility of the plastic membrane. The ingredients of the alloy vary from manufacturer to manufacturer, but the key factor is conductivity—it must be able to carry an electrical charge.

The metal-coated plastic will vibrate when it's subjected to an audio wave because of the physical reality called sympathetic vibration. The principle of sympathetic vibration says that if it is possible for a surface to vibrate at a specific frequency, it will vibrate when it is in the presence of a sound wave containing that frequency. The metal-coated plastic membrane, the diaphragm, in a condenser microphone must be able to sympathetically vibrate when in the presence of any audio wave in our audible frequency spectrum.

This metal-coated piece of plastic is positioned close, and parallel, to a solid piece of metallic alloy called a backplate. When a voltage is applied across the plates, an equal amount of electric charge is formed on each plate (positive charge on one plate, negative on the other). The air between the diaphragm and the back-

Microphones & Mixers.. by BILL GIBSON

plate is called the dielectric. It interrupts, but is influential in, the storage of electrons in the circuit.

As the crest and trough of a sound wave meet the thinly coated plastic, the plastic vibrates sympathetically with the sound wave. As the diaphragm vibrates, the distance between the solid metal surface and the moveable metal surface changes. These changes in the air gap create a variance in the storage of the electrical current. This current, made up of the charges moving on and off of the plates, parallels the changing energy in the sound wave. In other words, an extremely accurate electrical representation of the source acoustical sound wave is created.

Because there is very little mass in the condenser microphone's metal-coated membrane, it responds very quickly and accurately when in the presence of sound. Therefore, the condenser capsule is very efficient at capturing sounds with high transient content, as well as sounds with interesting complexities.

The signal that comes from the capsule is very weak and must be amplified to mic level. Most condenser mics use transistors in the internal amplifying circuitry; transistors provide a very clean and accurate amplification. Some condenser microphones utilize a vacuum tube instead of the transistor because of the

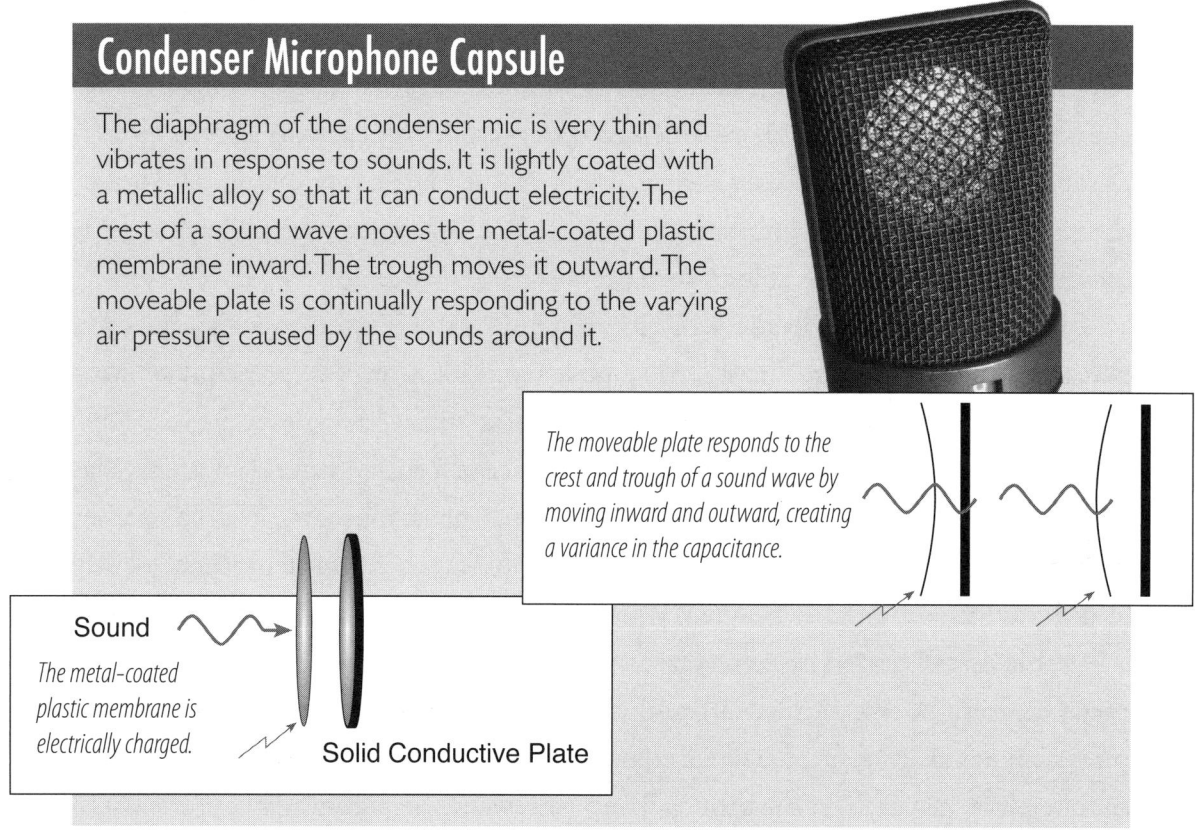

Condenser Microphone Capsule

The diaphragm of the condenser mic is very thin and vibrates in response to sounds. It is lightly coated with a metallic alloy so that it can conduct electricity. The crest of a sound wave moves the metal-coated plastic membrane inward. The trough moves it outward. The moveable plate is continually responding to the varying air pressure caused by the sounds around it.

The moveable plate responds to the crest and trough of a sound wave by moving inward and outward, creating a variance in the capacitance.

Sound

The metal-coated plastic membrane is electrically charged.

Solid Conductive Plate

Powering the Condenser Microphone Capsule

Condenser microphones are different from the other common mic types because they require power to operate the internal amplification circuit and to provide power to the mic capsule itself. This power is supplied by a battery, phantom power from the mixer, or from a power supply designed specifically for the mic being used.

When positive and negative terminals (left) are shorted together, current flows virtually unimpeded, heating up and burning the power supply out.

The dielectric interrupts the short circuit. The diagram to the right represents a simplified capacitor (condenser). The electrons stored in the system are influenced by changes in the distance between the plates (vibrations of the diaphragm).

From the power supply, the positive terminal is connected to the moveable plate (diaphragm), and the negative terminal is connected to the backplate. If the diaphragm and backplate of the condenser capsule were connected together, the battery or power supply would short out (heat up, explode, catch on fire, and so on). The distance between the plates determines the amount of electrical charge that is stored on each plate. When sound waves vibrate the diaphragm, the motion causes a variation in the distance between the plates. This, in turn, influences the electrical charges to move on and off of the plates, ideally in direct proportion to the air-pressure variations in the sound wave. The electric current caused by the varying capacitance is the audio signal.

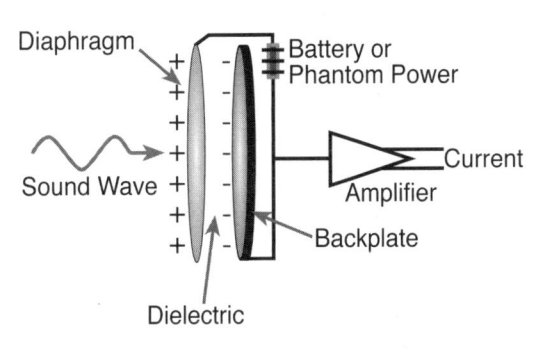

smooth and warm sound it produces. Many of these tube microphones are well respected and highly acclaimed.

Phantom Power

The capsule of a condenser microphone requires power to charge the metal-coated membrane. Power is also required to amplify the signal from the capsule up to microphone level.

Some condenser microphones are designed to be able to house a battery for powering the capsule and amplifying circuitry. However, phantom power provides a more efficient way to get power to the condenser mic because it's efficient,

Microphones & Mixers .... by BILL GIBSON

constant, and reliable. The phantom power supply is typically in the mixer. The power is sent to the mic through the balanced mic cable.

If you use batteries to power a condenser mic, always be sure the batteries are fresh and that they're supplying sufficient voltage to optimally run the microphone's circuitry.

If your mixer doesn't provide phantom power, use a commercially manufactured external phantom power supply. It receives 120-volt AC current and transforms it to the proper DC voltage and amperage. External mic preamplifiers also supply phantom power.

Phantom power voltage is typically 48 volts, although it can range from around 11 volts to 48 volts. Although low in voltage and amperage, a poorly designed or defective phantom power supply can be the source of electrical danger to the user. Each condenser mic draws current from the phantom power supply based on its electrical needs.

Voltage is the actual electro-motive force behind current. Current is measured in units called amps. A milliamp is one thousandth of an amp. Condenser mics draw a very low-amperage DC current, ranging from less than 1 mA (less than one one-thousandth of an amp) to about 12 mA. By comparison, a typical household circuit carries 15 to 50 amps of 120-volt AC current.

Phantom power has no adverse effect on the audio signal being carried by the mic cable. The DC voltage is applied equally to pins 2 and 3 of the XLR connection relative to pin 1, which is at ground potential. The fact that it functions undetected in the background on the same cable the mic signal travels on explains the term "phantom" power.

Vintage tube mics often don't require phantom power from the mixer because the power supply is external to the mic. The external power supply receives 120-volt AC current, which provides power to the external amplifying circuitry; the charging voltage for the capsule element is provided by the external power supply. The mic connects to the power supply, and then the power supply connects to the mixer input.

Electret Condenser Microphones

An electret condenser microphone utilizes a permanently charged capsule, which doesn't require phantom power. However, power is still required to operate the internal preamp. Phantom power can still be used to power the microphone, but the decreased electrical requirements make this condenser mic efficient while receiving battery power. Consequently, electret condensers are an excellent choice for application in the field. They possess all the sonic benefits of the condenser design with a realistic expectation that the battery power will provide sufficient longevity.

COMPARISON BETWEEN MOVING-COIL, RIBBON, AND CONDENSER MICROPHONES

If you possess the basic understanding of each mic type and how it works, you'll be able to make very intelligent microphone choices. The microphone you select for your specific recording situation makes a big difference to the sound of the final recording. It's amazing how easy it is to get great sounds when you've selected the right mic for the job and you've run the mic through a high-quality preamp.

Whereas the diaphragm of the microphone is the vibrating membrane that responds to sound waves, its makeup plays a key role in the inherent ability of the microphone to provide an accurate version of the sound it receives. Since we know that the moving-coil capsule utilizes a membrane attached to the top of a coil of copper wire, and since the sound wave must move the entire assembly around a magnet, we can draw the simple deduction that, by nature of its mass, it is physically less responsive than either the ribbon or condenser capsule. In fact, this deduction is true. Condenser mics are the most accurate and responsive of the three mic types, and ribbon mics are typically more accurate than moving-coil dynamic mics. Though there may be anomalies to this comparison, it is generally accepted.

MOVING-COIL MICS

Though moving-coil mics don't excel in capturing transients and subtleties, you can still take advantage of their tendencies and characteristics.

Moving-coil mics are the standard choice for most live situations, but they are also very useful in the studio. Here are some examples of popular and trustworthy moving-coil microphones:

+ AKG D12, D112, D3500, D1000E
+ Audio-Technica ATM25, Pro-25

Microphones & Mixers .... by BILL GIBSON

+ Beyer M88
+ Electro-Voice RE20
+ Sennheiser 421, 441
+ Shure SM57, SM58, SM7

Moving-coil mics are the most durable of all the mic types. They also withstand the most volume before they distort within their own circuitry.

A moving-coil mic typically colors a sound more than a condenser mic. This coloration usually falls in the frequency range between about 5 kHz and 10 kHz. As long as we realize that this coloration is present, we can use it to our advantage.

Moving-Coil Dynamic Microphones

These are some workhorse moving-coil microphones. They are rugged, durable, and with little or no maintenance, they are likely to still be functioning well in 30 years. These mics are typically best-suited to close-miking applications.

Shure SM58

This mic, along with the SM 57, might be the most commonly used mic of all time. It is rugged, well-designed, long-lasting, and sounds clean and full in almost any close-miking application.

Shure Beta 57

The Beta series updates and augments the popular SM series. These mics sound a little cleaner and crisper than their predecessors and they are still full and clean in the low end. They work well when close-miking electric guitars, drums, and live vocals.

Sennheiser MD421

The MD 421 is many engineers' favorite tom mic. Others prefer it for kick drum or electric bass. It has a characteristic sound that compliments the SM and Beta series mics from Shure.

Shure Beta 52

This large-diaphragm moving-coil mic quickly became the standard for miking kick drum and electric bass cabinets. It provides a full sound with a clean high-frequency range.

Electro-Voice RE20

An extremely popular broadcast announcer mic, the RE20 also sounds great on kick drum, electric bass cabinet, and other sounds that are rich in lows. It provides a clear high end and relatively flat frequency response.

Shure SM7

This is a great-sounding broadcast mic but it became most notable when veteran engineer, Bruce Swedien, used it to record Michael Jackson's vocals on the best-selling album of all time, Thriller.

In our studies on EQ, we've found that this frequency range can add clarity, presence, and understandability to many vocal and instrumental sounds.

Moving-coil mics have a thin sound when they are more than about a foot from the sound source. They're usually used in close-miking applications, with the mic placed anywhere from less than an inch from the sound source up to about 12 inches from the sound source.

Since moving-coil mics can withstand a lot of volume, they sound the best in close-miking applications; and since they add high-frequency edge, they're good choices for miking electric guitar speaker cabinets, bass drum, snare drum, toms, or any loud instrument that benefits from close-miking technique. Use them when you want to capture lots of sound with lots of edge from a close distance and aren't as concerned about subtle nuance and literal accuracy of the original waveform.

Moving-coils are also used in live performances for vocals. They work well in close-miking situations, add high-frequency clarity, and are very durable.

CONDENSER MICROPHONES

Condenser microphones are the most accurate. They respond to fast attacks and transients more precisely than other types, and they typically add the least amount of tonal coloration. The large vocal mics used in professional recording studios are usually examples of condenser mics. Condenser mics also come in much smaller sizes and interesting shapes. Some popular condenser mics are

- AKG 414, 451, 391, 535, C1000, 460, C3000, C-12, The Tube, Perception 820 Tube
- Audio-Technica 4033, 4050, 4047SV, 4047MP, 4060, 4041, 8022
- B&K 4011
- Blue Microphones Bluebird, Bottle, Cactus, Kiwi, Mouse, Dragon Fly, Blueberry, Baby Bottle
- Crown PZM-30D
- Electro-Voice BK-1
- Milab DC96B
- Neumann U87, U89, U47, U67, TLM170, KM83, KM184, TLM193
- Schoeps CMC 5U
- Sennheiser MKH 40, MKH 80
- Shure KSM 44, KSM 32, KSM 27, KSM 141, KSM 109 , SM 82

198

Microphones & Mixers .... by BILL GIBSON

Condenser Microphones

These microphones provide accurate and appealing recordings. A combination of solid-state and tube condenser mics is pictured here. In general, solid-state mics are the most accurate and the tube mics sound warmer and smoother. In addition, solid-state distortion radically clips the audio waveform, whereas tube distortion round off the peaks of overdriven signals.

Neumann U87

The U87, a solid-state condenser microphone, is one of the most widely used large-diaphragm studio condenser mics. It is a true classic and it sounds especially great on vocals and acoustic instruments.

Neumann U67

The U67, a tube condenser microphone, provides an amazingly warm and clear sound. It is a great choice for vocals and acoustic instruments. Like the U87, this is a classic recording tool that is a first-choice mic for many applications.

AKG C 451

This small-diaphgram condenser mic has been a staple in the recording engineer's toolkit for years. It is accurate, extremely flat, versatile, and very responsive to transients. This series contains interchangeable capsules for omni or cardioid pickup patterns.

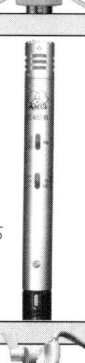

Audio-Technica 4041

Like the AKG 451, this small-diaphragm condenser mic responds well to transients and is very accurate. This class of mic records any percussive or acoustic instrument extremely well and sounds full and natural in distant-miking applications.

Neumann U47

This definitive classic vocal tube mic has been used by the biggest stars on some of the all-time best-selling recordings. It has warmth, personality, and it looks great. Its large diaphragm sounds smooth, clear, and intimate.

Audio-Technica 4047MP

The cardioid AT4047SV was built to match the vintage sound of Neumann's solid-state U47fet. It was immediately accepted and put to work. The AT4047MP is the multipattern version and also sounds fantastic.

AKG 820 Tube

The 820, part of the Perception series from AKG, is a large-diaphragm, affordable, and great-sounding tube mic. It uses a nine-position pattern selector located on the remote power supply for adjustment by the engineer when crafting the sound.

Audio-Technica 2020 USB

This new class of microphone connects to the computer via USB, providing an affordable, entry-level way for the home recordist to make better-sounding recordings. It sounds good and is very simple to connect.

Use a condenser microphone whenever you want to accurately capture the true sound of a voice or instrument. Condensers are almost always preferred when recording:

+ Acoustic guitar
+ Acoustic piano
+ Vocals
+ Real brass
+ Real strings
+ Woodwinds
+ Percussion
+ Acoustic room ambience

Condenser microphones typically capture a broader range of frequencies from a greater distance than the other mic types, especially in omnidirectional configuration. In other words, you don't need to be as close to the sound source to get a full sound. This trait of condenser microphones is a great advantage in the recording studio because it enables us to record a full sound while still including some of the natural ambience in a room. The further the mic is from the sound source, the more influential the ambience is on the recorded sound.

Condenser microphones that work wonderfully in the studio often provide poor results in a live sound reinforcement situation. Since they have a flat frequency response, these condenser mics tend to feed back more quickly than microphones designed specifically for live sound applications (especially in the low-frequency range). There are many condenser mics designed for sound reinforcement, and there are many condenser mics that work very well in either setting. Condenser mics often have a low-frequency roll-off switch that lets you decrease low-frequency sensitivity. In a live audio situation, the low-frequency roll-off is effective in reducing low-frequency feedback.

RIBBON MICS

Ribbon mics are the most fragile of all the mic types. This one factor makes them less useful in a live sound reinforcement application, even though ribbon mics produced within the last 10 or 15 years are much more durable than the older classic ribbon mics. It's still important to handle ribbon microphones carefully whether they're used in a live or studio application. Royer has developed a more rugged

Ribbon Microphones

From the early days of recording, ribbon microphones have been important. For a time, their popularity dwindled; however, with the advent and popularization of digital recording, enthusiasm for the warmth and smoothness provided by a ribbon microphone was reinvigorated. Beyer Dynamics provided a few ribbon microphone options throughout the '70s and '80s, but when Royer Labs brought their new-technology figure-of-eight ribbon microphones to market they were immediately embraced by professionals who missed the warmth and depth of the analog recording process. In recent years Audio-Technica, Shure, and AEA have also begun to provide exceptional ribbon microphones.

Royer R-121

When Royer released the R-121, it immediately caught the attention of the recording community. It provided clarity without edge and warmth without mud. Its full, smooth sonic signature was a great fit for recording brass, string, drums, and even electric guitar.

Royer SF24

The SF24, a very natural-sounding phantom powered stereo ribbon microphone, provides a very realistic and full stereo image. The shock mount lets the engineer orient the mic for either X-Y or M-S stereo recordings.

Beyer Dynamic M160

In the '80s, this was my introduction to modern ribbon technology. I was instantly drawn to the sound it provided on brass and it was the best mic I'd heard on snare drum. However, it was fragile, and after breaking two of them in one drum session, I went back to my other choices.

Royer R-122

The R-122 is similar to the R-121 except it utilizes phantom power and has a slightly more controlled low frequency response and a quicker transient response. It's a bidirectional, side address ribbon mic with exceptional clarity and warmth.

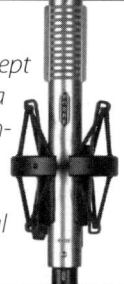

Royer R-101

The R-101 is a newer offering from Royer. It can withstand increased low frequency energy and has a more effective built-in windscreen, but it still has the characteristic warmth and depth expected from a ribbon mic.

Audio-Technica 4080

Audio Technica entered the ribbon market relatively recently and has provided some very nice mics. The AT4080 is phantom powered and implements new technologies to increase durability and sonic quality.

Audio-Technica 4081

The AT4081 is in a smaller physical housing than the AT4080, which makes it easier to position when miking certain instruments. It share the new ribbon technology incorporated in the 4080 and is a great-sounding microphone.

AEA R44

This recreation of the vintage RCA 44BX from the late 1930s is a fantastic ribbon mic. It provides nearly identical warmth and smoothness as the original, and it relies on new technology for increased durability and consistency.

series of ribbon mics specifically for live applications. They've increased the thickness of the ribbon from 2.5 microns in the studio version to 4 microns for the live version. This thicker ribbon increases durability though it slightly limits the high-frequency response. In live sound reinforcement applications, the loss of high-frequency response is unnoticed—a worthwhile trade-off for increased durability while retaining the ribbon microphone's characteristic warmth and tone.

The ribbon capsule is inherently bidirectional. Both the front and back of the ribbon are equally sensitive, and sound from the 90 and 270 degrees off-axis cancels. Many manufacturers take advantage of this natural characteristic and produce bidirectional ribbon mics. On the other hand, there are several ribbon mics that exhibit a unidirectional characteristic; these mics utilize a ribbon with the back (180 degrees off-axis) enclosed. Once the back of the ribbon is enclosed the capsule is inherently omnidirectional (like the moving-coil and single-diaphragm condenser capsules) and the enclosure determines the microphone's directional characteristic.

Ribbon mics, exhibiting a unidirectional polar characteristic, are like moving-coil mics in that they color the sound source by adding a high-frequency edge, and they generally have a thin sound when used in a distant miking setup. When used as a close mic, ribbon microphones can have a full sound that is often described as being warmer and smoother than a moving-coil.

There are some great-sounding ribbon microphones available. Some of the commonly used ribbon microphones are:

+ AEA R84, R44C, R44CX, R88
+ Audio-Technica 4080, 4081
+ Beyer M160, M500
+ Coles 4038
+ RCA 77DX, 44BX, 10001
+ Royer R-101, R-121, R-122, SF-12, SF-24
+ Shure KSM 313, KSM 353

Ribbon mics are fragile and need to be used in situations where they won't be dropped or jostled. If you use a ribbon mic to record drums and the drummer hits the mic too many times with his stick, the ribbon will break. Repairs like this can be costly. After breaking a couple of these mics, I decided it might be best if I stuck to one of the tried and true, very durable choices. I still tend to use the Beyer M160 ribbon a lot when I'm recording drum samples because I like the sound,

Microphones & Mixers... by BILL GIBSON

but sampling is a very controlled mic usage, and I'm usually the only one around with a drum stick.

SHAPING THE PICKUP PATTERN

The inherent polar response characteristic of the moving-coil and basic condenser capsules is omnidirectional. Set in space with no physical housing, they are equally sensitive to sound from all directions—they don't reject sound from any direction. The inherent polar response characteristic of the ribbon capsule is bidirectional. Set in space, with no physical housing, it is equally sensitive to sound from the front and back but it is not sensitive to sounds coming from the sides—it rejects sounds 90 and 270 degrees off-axis.

Whereas the moving-coil and condenser capsules are inherently omnidirectional, and the ribbon capsule is inherently bidirectional, in design and application the majority of microphones ever manufactured exhibit a cardioid polar response. Polar response characteristics are designed and created in two fundamental ways: physical housing design and electrical combinations of multiple capsules.

Physical Housing Design

Most microphone designs use the physical housing around the capsule to shape directional characteristics. The concept is simple once you understand phase interactions between sound waves—amplitudes in phase sum, and amplitudes out of phase cancel. For physical shaping of directional characteristic, the ribbon capsule is omnidirectional because the back of the ribbon is enclosed.

Microphones that contain a capsule at the top of a barrel housing with slots (we could also say openings, or ports) around the capsule end exhibit a unidirectional polar response. It is these ports that shape the directional characteristic. On-axis sound waves stimulate the diaphragm in the normal way; however, off-axis sound waves are allowed to reach the front and/or back of the diaphragm via multiple routes, provided by the ports around the housing. As the off-axis waveforms combine at the diaphragm, the fact that they've arrived at the same point (the diaphragm) through various pathways (around the mic, and through the network of ports) indicates that the length of their journey varies with the pathway. Since the same off-axis waveform has been split, and since each pathway is a different distance from the origination of the waveform, phase interaction is

built into the design. The key in the mic design is the positioning and quantity of the ports. The intent of the designer is to allow negative phase interaction of off-axis waveforms, through the series of ports in the physical housing, as they travel multiple off-axis pathways.

If the ports are covered up, these microphones become omnidirectional rather than cardioid. If fact, some manufacturers offer multiple capsules for use with the same microphone body. The capsules typically screw on and off and their only physical difference is that the cardioid capsule has ports and the omni capsule doesn't.

· Audio Example 5-2

Acoustic Guitar Miked Using Multiple Polar Patterns

The large format condenser microphone controls directional characteristic electrically rather than through the design of the physical housing. These microphones

Creating the Cardioid Pickup Pattern

A microphone with selectable pickup patterns typically uses a variance in relative polarizing voltage between the front and back sides of the condenser capsule to shape the pattern between omnidirectional, bidirectional, and variations of the cardioid patterns. Other mic designs utilize the physical design of the mic housing to influence the polar pattern.

The ports on the sides of the housing provide an alternate pathway to the mic capsule for off-axis sounds. As sounds travel around the mic, through the ports, and arrive at the front and back of the diaphragm, they reduce in level because of phase cancellation. The position and quantity of the ports determines the specific frequencies that are rejected most.

The microphone to the right is a Shure KSM141. It switches from cardioid to omnidirectional characteristic by sliding an internal cylinder up to cover the ports.

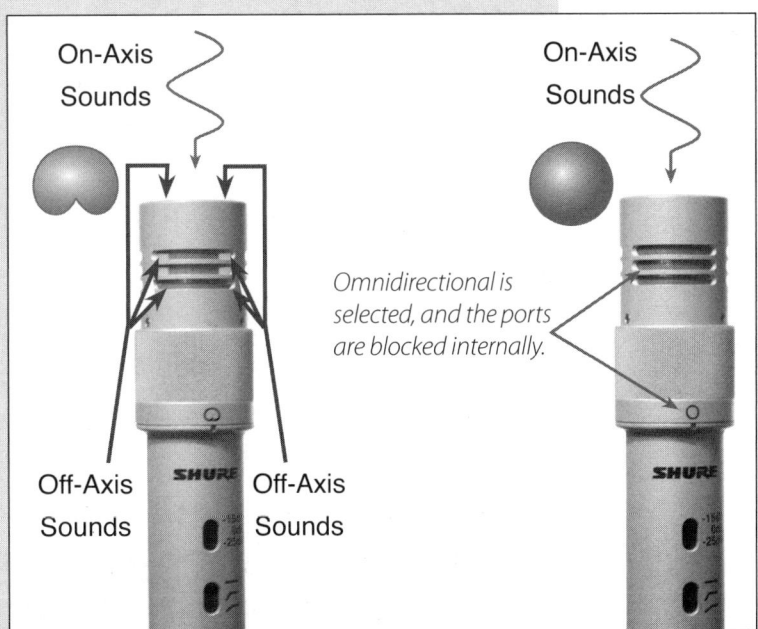

204

Microphones & Mixers.. by BILL GIBSON

Creating Polar Patterns Electrically

One moveable plate is on each side of the solid, fixed plate. Sound is received on both sides of the capsule.

Positive or negative charge

Large-diaphragm studio condenser mics with selectable patterns utilize a double-sided capsule with two moveable plates. The plates are charged with positive or negative polarity, in varying amounts, to shape virtually any polar response pattern. Some of these mics actually offer external control over polar response, which infinitely varies between patterns. The engineer, in the control room, shapes the microphone response to match the room and the source.

Selectable Polar Patterns

Polarity requirements for each polar response pattern:

*   Applying a positive charge to both moveable plates produces an omnidirectional response characteristic for the capsule.

*   Applying a positive charge to one plate and a negative charge to the other produces a bidirectional pattern.

*   Varying the relative intensity of the charge between the two plates, as well as changing the backside plate from positive to negative, produces any variation or permutation of cardioid, bidirectional, and omnidirectional characteristics.

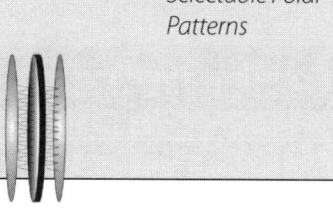

+	+	= omnidirectional
+	−	= bidirectional
+	−	= cardioid patterns
+	+	= cardioid patterns

are designed for and used in the studio and are rarely seen in a live setting; they're large, expensive, fragile, and adversely affected by environmental conditions like humidity, smoke, wind, etc.

Many of these microphones provide multiple pickup patterns, selectable by a switch on the mic or, externally, on a remote control. Selecting different pickup patterns causes a change in electrical polarization; nothing is altered as far as the physical housing is concerned. These microphones achieve multiple directional characteristics by incorporating a double-sided capsule. This is merely an extension of the condenser capsule we discussed earlier; however, in this design there are two moveable plates, one on either side of the fixed backplate.

In this design, the microphone's directional characteristic is controlled through the application of varying amounts of positive and negative electrical charges to the two moveable plates:

*   When both plates receive a positive charge, the mic exhibits omnidirectional characteristics.
*   When the front plate receives a positive charge and the back receives a negative charge, the mic becomes bidirectional.
*   As the intensity of the charging voltage varies between the two plates, the pickup pattern can be shaped at will, ranging from omnidirectional to bidirectional to cardioid, and on many mics multiple patterns in between. Most of these mics let the user select between preset pickup patterns through the use of a switch on the front of the mic. A few manufacturers offer a remote control for pattern selection, with some even offering a continuously variable balance of polarizing voltage. With this control, you can evaluate the sound in the control room, capturing the balance of the direct and ambient sound that makes the most musical impact.

·· Video Example 5-2

Demonstration of Polar Pattern Changes

Some microphone designs actually incorporate two capsules to create directionality. An omnidirectional pickup pattern combined with a bidirectional pickup pattern produces a cardioid response.

THE PROXIMITY EFFECT

It's important to understand the proximity effect. The result of the proximity effect is this: As the microphone gets closer to its intended source, the low frequency range increases in relation to the high frequencies. It's not uncommon to see a rise of 20 dB at 100 Hz as the source gets close to the mic. Anyone who has used a handheld mic has probably recognized that, if they get closer to the microphone, the sound becomes bigger, louder, and more bass-heavy—this happens as a result of the proximity effect.

The proximity effect is most pronounced when using a microphone with a cardioid or bidirectional pickup pattern; it is least pronounced when using a microphone with an omnidirectional pickup pattern.

Microphones & Mixers.. by BILL GIBSON

In effect, as a result of the proximity effect, the low frequency range increases in relation to the high frequency range. In reality, as the microphone moves closer to the source—a face, for instance—reflections from the source reflect back to the capsule out of phase, canceling more and more of the upper frequencies. The reflections not only cancel at the diaphragm, but they also enter through the ports, from behind the capsule, and cancel. This explains why there is less of a problem with proximity effect when using a omnidirectional microphone—there's less cancellation because there are no ports.

.... Video Example 5-3
Demonstration of the Proximity Effect

Compensating for the Proximity Effect

To help compensate for the proximity effect, many microphones have a user-selectable highpass filter built in. When you're close-miking anything where the sound becomes too bass-heavy, simply apply the highpass filter. A typically high-pass filter sets the cutoff frequency at 75 or 80 Hz. Some mics even let the user determine the cutoff frequency, typically offering various choices between 60 and 250 Hz.

Using the Proximity Effect to Our Advantage

The proximity effect isn't necessarily a bad characteristic of a microphone design, especially when we consider the effect it has on the close-miked sound in relation to the microphone's frequency response characteristic. Often, microphones with a unidirectional pickup characteristic exhibit a decreased sensitivity in the low frequencies, especially in a distant-miking application, and they're likely to exhibit an increased sensitivity in the high frequencies between 4 and 8 kHz.

Once we understand these tendencies, we realize that many microphones used in a close-miking application benefit from close proximity positioning. The increased low frequencies fill out an otherwise thin sound, and the extra sensitivity in the high-frequency range helps clean up the sound, resulting in greater understandability, presence, and clarity.

RESPONSE CHARACTERISTIC

Almost any microphone responds to all frequencies we can hear, along with frequencies above and below what we can hear. The human ear has a typical frequency response range of about 20 Hz to 20 kHz. Some folks have high-frequency hearing loss, so they might not hear sound waves all the way up to 20 kHz, and some small children might be able to hear sounds well above 20 kHz.

Frequency Response Curve

For a manufacturer to tell us that their microphone has a frequency range of 20 Hz to 20 kHz tells us absolutely nothing until they tell us how the mic responds throughout that frequency range. A mic might respond very well to 500 Hz, yet it might not respond very well at all to frequencies above about 10 kHz. If that were the case, the sound we captured to tape with that mic would be severely colored.

We use a frequency response curve to indicate exactly how a specific microphone responds to the frequencies across the audible spectrum.

The frequency response curve is the line on the graph that indicates the microphone's ability to reproduce frequencies across the audible spectrum. As the

Frequency Response Curve

A mic with a flat frequency response adds very little coloration to the sound it picks up. Many condenser microphones have a flat, or nearly flat, frequency response. This characteristic, combined with the fact that they respond very well to transients, makes condenser mics very accurate.

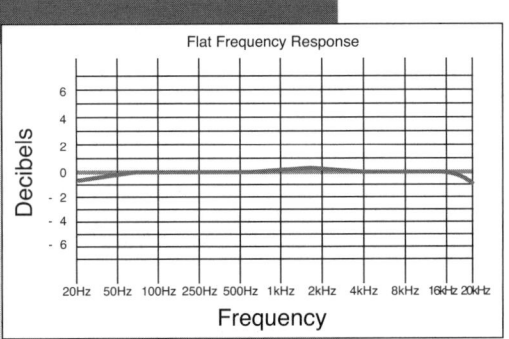

The mic represented by the curve at the right isn't very good at recording low frequencies and it produces an abundance of signal at about 4 kHz. Though this mic wouldn't be very accurate, we could intelligently use a mic like this if we wanted to record a sound with a brutal presence. Many moving-coil microphones have this kind of frequency response curve. Moving closer to the mic helps fill out the low frequencies.

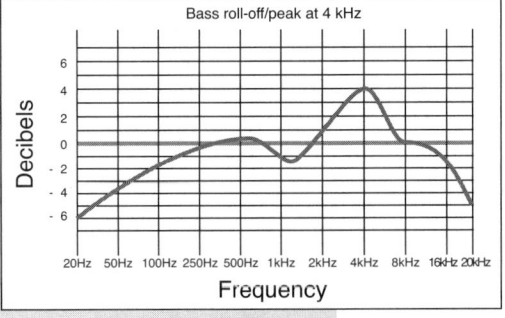

Microphones & Mixers.. by BILL GIBSON

sensitivity to a frequency increases the curve ascends; as the sensitivity decreases the curve descends.

A microphone that is equally sensitive to all frequencies across the audible spectrum is represented by a flat-line curve on the 0 dB line—this is called a flat frequency response.

If a frequency response curve shows a peak at 5 kHz, we can expect that the mic will color the sound in the highs, likely producing a sound that has a little more aggressive sound than if a mic with a flat response was used. If the frequency response curve shows the low frequencies dropping off sharply below 300 Hz, we can expect the mic to sound thin in the low end unless we move it close to the sound source to proportionally increase the lows.

A frequency response graph often contains more than one curve. The main, and typically more solid and bold curve, represents the on-axis frequency response. Other curves on the graph represent off-axis frequency response and, sometimes, comparisons between the responses in multiple sound fields.

TRANSIENT RESPONSE

The frequency response curve is a valuable tool to help us predict how a mic will sound. However, what the frequency response curve doesn't tell us is how the mic responds to transients. We can predict the transient response of a mic based on what we already know about the basic operating principles of the different mic types. Therefore, condenser mics are expected to more accurately capture the fast transient, with moving-coil and ribbon mics lagging behind.

It's important to note that all commonly used studio microphones do the same basic task: they respond to changes in air pressure and convert those changes in pressure to a continuously varying flow of electrons. We've just seen this in the operational principles of the moving-coil, condenser, and ribbon capsules. In addition, while transient response is a fundamentally important to the sound of a microphone, it is only one aspect of the microphone's response over time.

The way the capsule stops vibrating also impacts the sound and feel of the miked source. The mass of a moving-coil capsule slows down the rate of stabilization after an attack. Similarly though not as extreme, a large-diaphragm condenser capsule takes a little longer to settle down than a small-diaphragm condenser capsule. Microphones with small condenser capsules are, by design, the most accurate

Response Comparison

Below, there are 12 different waveforms from 12 different microphones, all subjected to the same woodblock, which was stuck with the butt end of a drumstick. Each group of four microphones was recorded into Pro Tools at the same time, capturing the exact same woodblock sound through identical mic preamplifiers. The EQ is flat and there are no effects. The visual waveforms represent very interesting differences in the microphones' ability to respond over time. Some ramp up quicker than others and some take a lot longer to ramp down and do so in a much different way. Mic specifications reveal interesting aspects of microphones and can help us predict how they might sound but there is no formal specification that defines what these images show. Keep in mind that there is not necessarily a better or worse mic because of these representations. The real test is in their ability to capture the desired sounds in the desired manner.

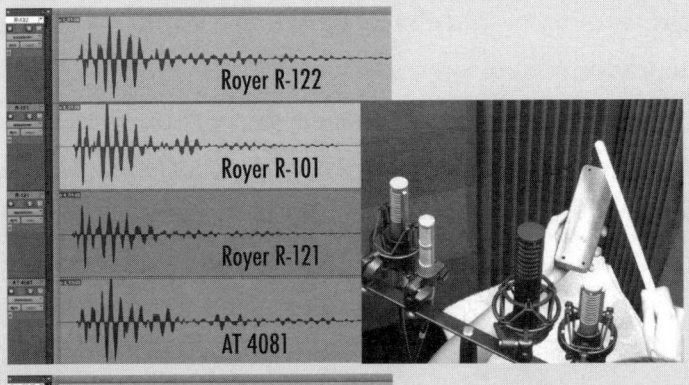

This group of four ribbon microphones all sound great in use but they're all very different in the way they respond along the time axis. It's not difficult to see that they each responding in a unique manner to the woodblock sound.

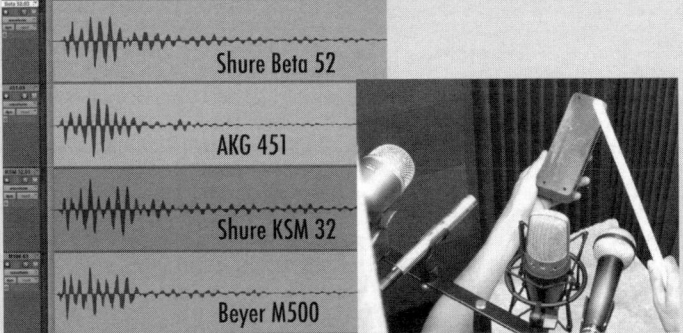

This is the most diverse group of microphones, including the Shure Beta 52 moving-coil, AKG 451 small-diaphragm condenser, Shure KSM 32 large-diaphragm condenser, and Beyer M500 ribbon. Their differences are extreme.

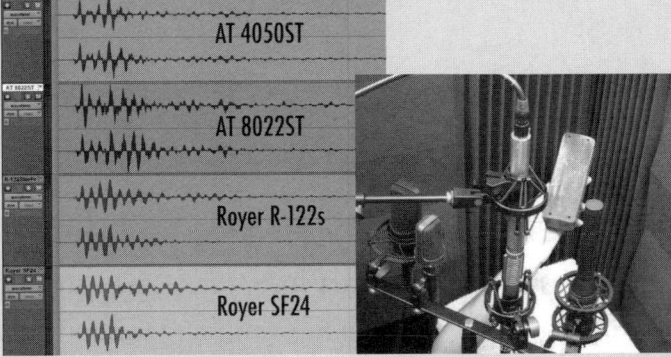

These are all wonderful stereo microphones but their waveforms, from the identical source, are very different. Be sure to watch and listen to Audio and Video Examples 5-1 to hear the differences between many of these mics.

Microphones & Mixers.. by BILL GIBSON

from the onset of the attack throughout the length of the sound and to the end of the decay. Microphones used for calibration and acoustical analysis use very small and accurate condenser capsules.

Interestingly, if all we wanted was accuracy, we'd expect to see most mics utilizing extremely small condenser capsules—that's just not the case. The capsules in common small-diaphragm condensers such as the Neumann KM 84, AKG 451, and AT 4041 are somewhat large relative to the capsule in mics used for calibration. The capsules in large-diaphragm condensers such as the Neumann U87, Audio-Technica 4047, and Shure KSM 44 are much larger still. The most important consideration is the way the microphone sounds. Recording music and other audio sources isn't typically about simply archiving an event—it's about building a work of art. That requires color and depth and character, which is what all of these different microphones so adeptly provide.

While there's no real specification that quantifies a microphone's transient response, an understanding of these basic design concepts can easily aid in making intelligent choices. It all comes down to listening and choosing the mic that fulfills your creative needs. Yes, there still is room for listening in this game of microphone choice!

Audio Example 5-3
Demonstration of Microphone Transient Response Characteristics

OUTPUT CHARACTERISTIC

The output characteristic of a microphone quantifies factors like noise, sensitivity, overload limits, and impedance.

Equivalent Noise Rating/Self-Noise

A microphone's equivalent noise level, also referred to as self-noise, indicates the sound pressure level that will create the same voltage as the noise from the microphone.

Moving-coil and ribbon microphones are very quiet. They contain passive circuitry, which poses very little noise potential, and they have a very low self-noise.

A condenser microphones produces more self-noise, simply because it contains amplifying circuitry necessary to boost the capsule signal up to mic level.

The amplifying circuitry, though adding to the mic's self-noise, provides a hotter signal to the mixer; therefore, it needs less preamplification at the mixer input. So, there's a bit of a trade-off in terms of cumulative noise at the mixer output. There's slightly more mic noise and slightly less mixer noise.

Sensitivity

We briefly covered microphone sensitivity rating in Chapter 3. To continue, in assessing sound energy, dB SPL refers to the energy or force a sound transfers into air—it's a measure of acoustical energy. The quietest sound the human ear can hear is considered to be 0 dB SPL and 1 dB SPL is considered the smallest increment of change in dB SPL that the human ear can recognize as a change. In order to fairly compare microphones we must understand how the manufacturers assess their products—they're not all the same. Microphone manufacturers must specify the acoustic input level at the microphone capsule.

Interestingly enough, there are two common references and they are based on how loud typical human speech is at distances of one inch and one foot. The sound energy created by human speech at a distance of one inch from the talker's mouth is considered to be 94 dB SPL, which is also the definition of one pascal. The intensity of human speech from a distance of one foot is considered to be 74 dB SPL, which is considered 0.1 pascal.

Microphone ratings are taken referenced to one of the previous energy levels and the exact reference will be in the microphone specifications. In addition, these measurements are taken without a load on the mic capsule, which means there is no cable connected to the microphone output end. This measures what is called the "open circuit voltage." A frequency of 1 kHz is considered an average reference-level tone, so that is typically the frequency applied to the mic capsule, either at 94 or 74 dB SPL.

The following example is from the specifications sheet of the Shure Beta 58A. It indicates a sensitivity of -51.5 when exposed to 1 pascal (94 dB SPL) at 1 kHz:

+ Output level (at 1,000 Hz)
+ Open circuit voltage: -51.5 dBV/Pa* (2.6 mV)
+ 1 pascal = 94 dB SPL
+ Impedance: rated impedance is 150 Ω (290 Ω actual) for connection to microphone

- Inputs rated low Z
- Phasing: positive pressure on diaphragm produces positive voltage on pin 2 with respect to pin 3

And this is an example of the spec sheet from the Audio-Technica 4047SV large diaphragm condenser microphone, which indicates a sensitivity of -35 dB:

- Open circuit sensitivity -35 dB (17.7 mV) re 1V at 1 pascal*
- Impedance 250 ohms
- Maximum input sound level 149 dB SPL, 1 kHz at 1% T.H.D.; 159 dB SPL with 10 dB pad (nominal)
- The asterisk after "pascal" refers to: *1 pascal = 10 dynes/cm2 = 10 micro-bars = 94 dB SPL

Notice that the Shure Beta 58A has a sensitivity of -51.5 dB and the AT 047 has a sensitivity of -35 dB. This means that the 4047 puts out a stronger signal in response to the same source because it requires less of a boost to achieve line level (35 dB), relative to the Beta 58A (51.5 dB). This is what we would expect.

In general:

- Condenser microphones are typically the most sensitive, with ratings in the range of -30 dB to -40 dB or so.
- Moving-coil microphones are next in line, with normal ratings in the -50 to -60 dB range.
- Ribbon microphones are often the least sensitive, with ratings between -58 to -60 or so.

These comparisons are general and historically correct, considering classic microphone designs. Technology provides for an extension of sensitivity in all the mic types. In fact, some of the modern ribbon mics actually receive phantom power in order to increase efficiency and to power an internal preamp much like the condenser mic.

Maximum SPL Rating

Most microphones can handle a lot of level (dB SPL) before they induce distortion. However, we sometimes need to mike loud instruments at close range, so distortion at the microphone can be an issue. It's important to be aware of the dB SPL where a specified percentage of distortion occurs (Total Harmonic Distortion). In addition, we need to realize the dB SPL, where the signal from the mic

will clip, is the maximum SPL rating referring to peak SPL. Keep in mind that the peak SPL rating is typically 20 dB greater than the average, or RMS, rating.

Moving-coil mics are capable of handling a bunch of level. They don't contain much other than the capsule and they're usually capable of handling peak SPL well in excess of 140 dB SPL peak, at acceptable distortion levels.

Condenser capsules are also capable of handling substantial level; however, the amplifying circuitry is likely to distort when subjected to loud sounds. Because of this, most condenser mics contain a pad, which decreases the signal strength from the mic capsule to its internal amplifying circuit (typically in 10-dB increments). Keep in mind that applying the pad diminishes the inherent signal-to-noise ratio by the amount of the pad, so implement it only when necessary.

Maximum SPL ratings always must be quantified at a specified total harmonic distortion (THD) and must be measured for the complete microphone (capsule and internal preamp). Most specifications relate maximum SPL to .05% THD, though some reference 1% THD.

The distortion of a circular capsule doubles with each 6-dB increase in level, so it's a simple matter to calculate distortion ratings in relation to the published specification. If a microphone specifies 140 dB SPL peak at 0.5% THD, it's implied that the same microphone will exhibit 1% THD at 146 dB SPL peak, or .25% THD at 134 dB SPL peak.

Impedance

In the modern recording world microphones are low-impedance devices. Most low-impedance mics fall in the impedance range between 50 and 250 ohms. High-impedance microphones are not common today; they were designed and optimized for use with vacuum tube amplification. High-impedance mics fall in the impedance range between 20,000 and 50,000 ohms.

Low-impedance ranges—mics with lower impedances, around 50 ohms—are more sensitive to electromagnetic hum and less susceptible to electrostatic interference, in relation to mics with impedances around 250 ohms. On the other hand, mics with impedances around 250 ohms are less sensitive to electromagnetic hum and more sensitive to electrostatic interference.

No matter what inherent noises mics with various impedance ratings are sensitive to, balanced low-impedance microphones are able to take advantage of

Microphones & Mixers .... by BILL GIBSON

the noise-canceling aspects of balanced circuitry; therefore, they benefit from the ability to run long cable lengths devoid of serious noise issues.

STEREO MIC TECHNIQUES

The use of stereo miking can provide wonderfully natural and powerful recordings. However, stereo techniques are often used on multiple voices, instruments, or other sound sources, and the control that is provided by close-miking is lost. Though there are a few changes that can be made tonally, certain portions of the sonic color and environmental ambience are permanent. In other words, be sure that you want the sound you're recording.

If there's a question about a certain sound and how it will actually work in the final mix, set up a close mic and record both the more-distant stereo setup and the close mic. When it's time to mix, you'll thank yourself for having the foresight to record both.

Let's examine a few of the standard stereo miking configurations. Each one of these techniques is field-tested—they have proven functional and effective. Listen very carefully and analytically to these examples. Listen for left/right positioning and for the perception of distance. Are the instruments close or far away? Can you hear a change in the tonal character as the different sounds change position? Do you perceive certain instruments as being above or below other instruments? Can you hear the room sound? In other words, pick these recordings apart bit by bit.

The X-Y Technique

The X-Y technique is one of the most commonly used stereo mic techniques. It works well in both live and studio applications. This technique uses two mics in a coincident physical relationship. Their capsules are as close together as possible, aimed across the same physical plane, and aimed at a 90-degree angle to each other. Listen very closely to the sound of each ingredient in the stereo recording. Listen to the changes in the sounds as they move around the room.

• Audio Example 5-4
X-Y Configuration

The X-Y technique can be extended by separating the two microphones. They can be moved so that the capsules are a few inches or a few feet apart, depending

X-Y Configuration

This is the most common stereo miking configuration. The fact that the microphone capsules are as close to the same horizontal and vertical axes as possible gives this configuration good stereo separation and imaging while also providing reliable summing to mono.

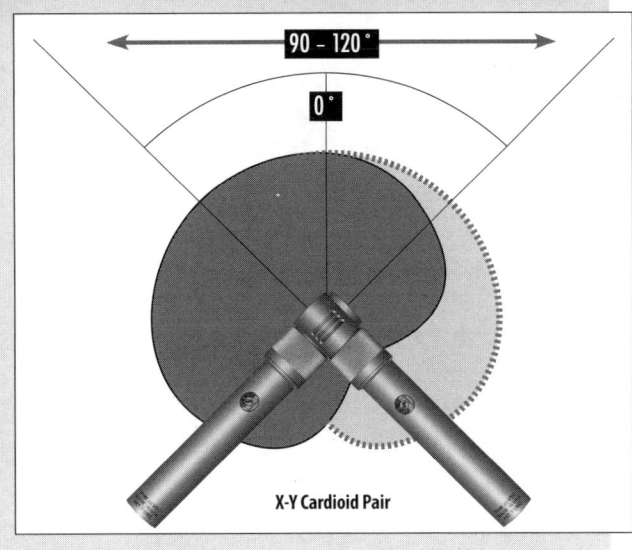

X-Y Cardioid Pair

on the width of the sound source. Always keep the mics angled about 90 degrees away from each other to maintain the stereo relationship. Never aim the two mics toward each other at the same source because the stereo image will be lost—two mics aimed at the same source functionally become a single mono mic with destructive phase interactions. Even though the X-Y pattern can be extended, the capsules should always remain along the same horizontal or vertical plane.

The Spaced Omni Pair

Two omnidirectional mics spaced between three and ten feet apart can produce a very good stereo image with good natural acoustic involvement. Using a spaced omni pair is primarily a recording technique because the omnidirectional microphones are prone to causing feedback in live applications.

When you are recording a small group, such as a vocal quartet, keep the mics about three feet apart; for larger groups increase the distance between the microphones. Use this technique only if the room has a good sound. In Audio Example 5-5, listen closely for the panning placement and perceived distance for each instrument. There's a definite difference in the apparent closeness of these percussion instruments.

Spaced Omni Pair

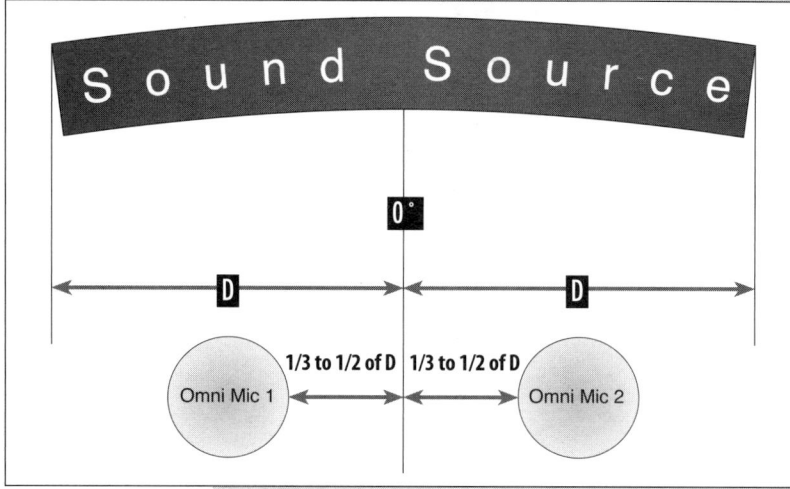

This configuration uses two omnidirectional mics. The ambience of the recording environment will color the sound of the recording. This setup is capable of capturing beautiful performances with great life—especially if the recording environment has an inherently good sound.

"D" on the diagram represents the distance from the center of the sound source to its outer edge. Notice that the distance from the center of the sound source to each microphone is one-third to one-half of D.

•••••••••••••••••••••••••••••••••Audio Example 5-5

Spaced Omni Pair

A variation of the spaced omni pair of mics involves positioning a baffle between the two mics, which increases the stereo separation and widens the image. This is primarily a recording technique because of the omnidirectional mics and

Spaced Omni Pair with Baffle

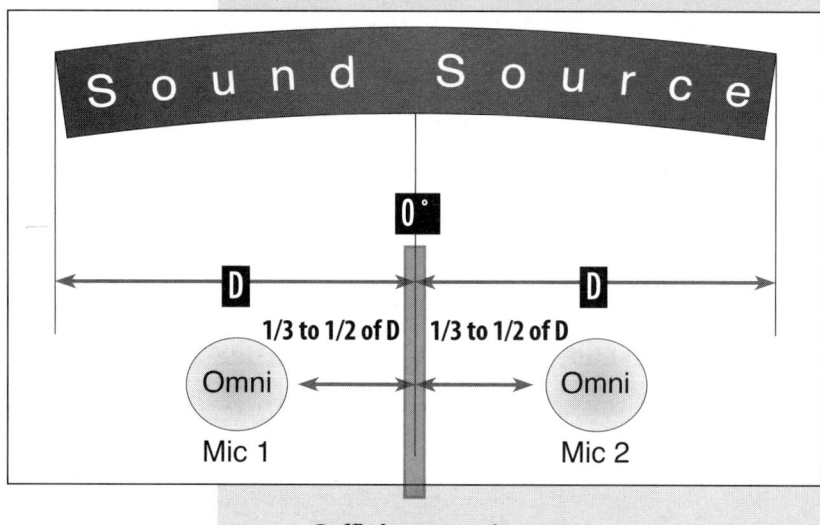

Baffle between mics

This technique retains much of the openness of the regularly spaced omni pair; however, the addition of a baffle between the microphones increases stereo separation. When miking a blended acoustical group or a stereo send from the multitrack of specific mix ingredients, this configuration provides a striking stereo image.

the awkwardness of positioning a baffle on stage. Notice, in Audio Example 5-6, how clearly defined the changes are as the percussion instruments move closer to and farther away from the mics.

·····························Audio Example 5-6

Spaced Omni Pair with a Baffle

The Crossed Bidirectional Blumlein Technique

The crossed bidirectional configuration Blumlein technique uses two bidirectional mics positioned along the same vertical axis and aimed 90–120 degrees apart along the horizontal axis. This is similar to the X-Y configuration in that it transfers well to mono, but the room plays a bigger part in the tonal character of the recording. This is primarily a recording technique.

·····························Audio Example 5-7

The Crossed Bidirectional Configuration

Crossed Bidirectional Blumlein Configuration

The crossed bidirectional configuration (also called the "Blumlein" configuration) has the advantage of being a coincident technique in that the overall sound isn't significantly degraded when the stereo pair is combined to mono. The sound produced by this technique is similar in separation to the X-Y configuration, but with a little more acoustical life.

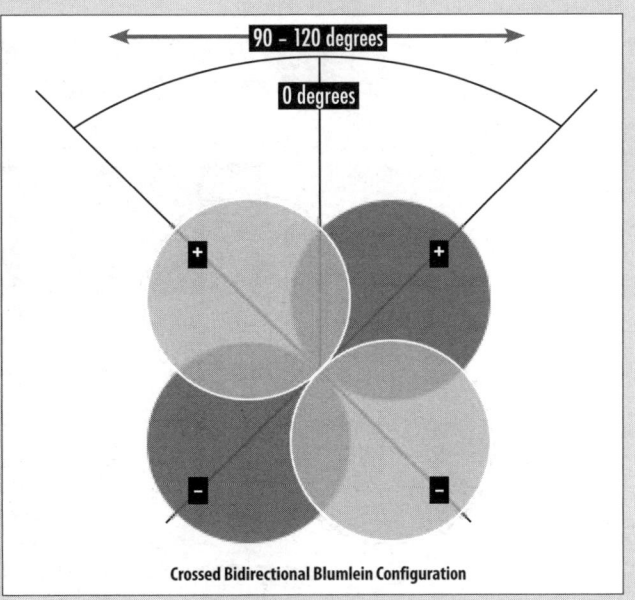

Crossed Bidirectional Blumlein Configuration

218

Microphones & Mixers .... by BILL GIBSON

The MS (Mid-Side) Technique

The MS technique is the most involved of the techniques, but it's the best in terms of combining stereo to mono, and it also gives a very true and reliable stereo image. Many stereo mics contain two condenser capsules that are positioned in an MS configuration.

Its only real drawback is that it isn't simple to hook up using two separate mics. You must use a combining matrix that'll facilitate sending the sum of the mid and side mics to one channel and the difference of the mid and side mics to the opposite channel. In other words, you must be able to:

1. Split or Y the output of the mid mic and send it to both channels (or simply pan it to the center position).

2. Split or Y the output of the side (bidirectional) mic and send it to both channels.

3. Invert the phase of one leg of the side mic split. A leg is simply one side of the Y from the side mic.

4. Leave the other leg of the side mic split in its normal phase.

5. Adjust the balance between the mid mic and the side mics to shape the stereo image to your taste and needs.

High-quality, double-capsule stereo mics typically use this configuration. They demonstrate the advantages of coincident technique—minimal phase confusion between the two microphones. Also, and possibly more important, since

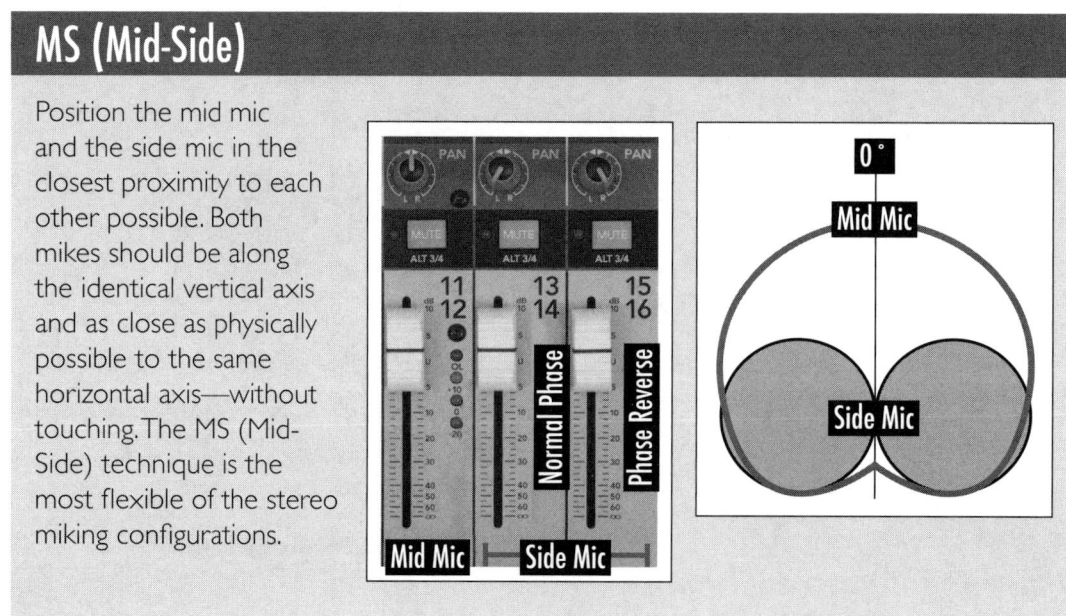

MS (Mid-Side)

Position the mid mic and the side mic in the closest proximity to each other possible. Both mikes should be along the identical vertical axis and as close as physically possible to the same horizontal axis—without touching. The MS (Mid-Side) technique is the most flexible of the stereo miking configurations.

MS Using Two Bidirectional Mics

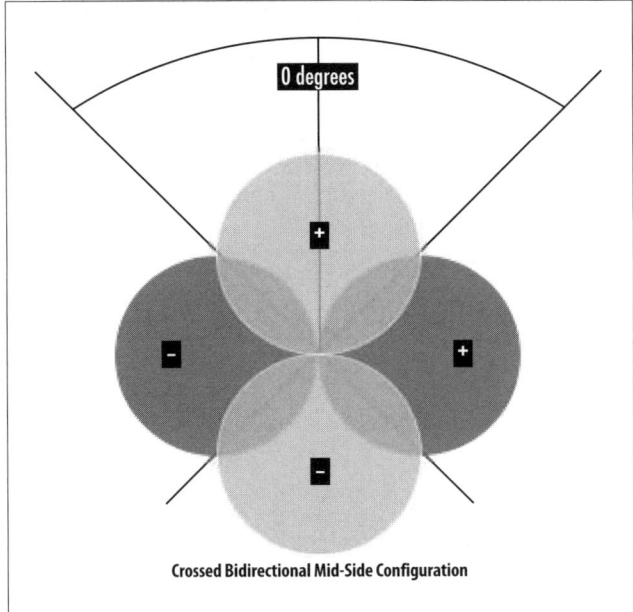

Crossed Bidirectional Mid-Side Configuration

The crossed bidirectional configuration can also be used to set up an MS technique. Some stereo ribbon microphones contain a switch that changes from an X-Y to an MS configuration. In this arrangement, the front of one bidirectional mic faces the source. The other mic faces 90 degrees to either side so that the side of the mic faces the source. The side mic is sent to one stereo channel and simultaneously to the other stereo channel 180 degrees out of phase. This provides a wider stereo image than the simple crossed X-Y bidirectional setup and it still collapses perfectly to mono.

the side mic signal is split to left and right—and left and right are made to be 180 degrees out of phase with each other—when the stereo signal is sent through a mono playback system, the side mic information totally cancels. This leaves the mid mic signal as simple and pure as if it were the only mic used.

. . . . Audio Example 5-8

Mid-Side Configuration

The ORTF Configuration

The ORTF technique was devised around 1960 at the Office de Radiodiffusion Télévision Française (ORTF) at Radio France. It is similar, in a way, to the X-Y technique and it looks like a spaced X-Y with the stereo pair of mics pointing away from each other. However, the angle between the two microphones is specified as 110 degrees and the distance between them is 17 centimeters. As with nearly all stereo mic techniques, the ORTF technique works best when the stereo pair is identical and matched for frequency consistency.

220

Microphones & Mixers... by BILL GIBSON

ORTF Configuration

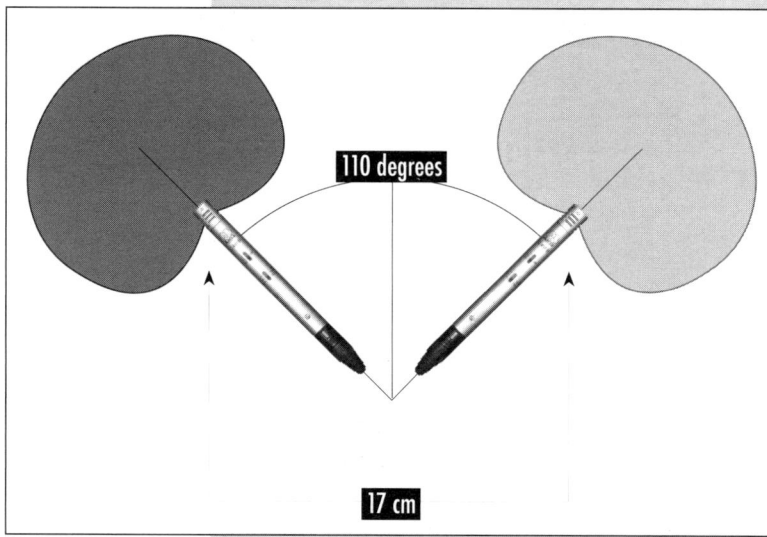

110 degrees

17 cm

The stereo mic technique developed by the Office de Radiodiffusion Télévision Française (ORTF) at Radio France provides a realistic stereo sound that offers a slightly wider image in comparison to the standard X-Y configuration. This technique produces excellent results when using two high-quality cardioid condenser microphones. The mics should be aimed 90–110 degrees apart and the distance between the two capsules should be 17 centimeters. Special mounting brackets help facilitate accurate and repeatable positioning of the mics for the ORTF configuration.

MULTICHANNEL MIC TECHNIQUES

As technology has facilitated the dramatic increase in data storage and playback capabilities, the world of multichannel audio has quickly evolved. Most DAWs provide at least some capacity to output all of the primary surround configurations.

As playback configurations standardize beyond the stereo limitations of the common CD, the modern recordist must become familiar with techniques that capture and reproduce multichannel imagery.

Simply pointing microphones in the directions from which you want to provide perspective to the listener can produce interesting and sometimes exemplary results, but there are pitfalls in a random approach that can cause problems. In most surround applications, the listener will feel most comfortable when presented with an aural image that feels like it could occur naturally. Techniques such as LRFB (left, right, front, back) and the use of specifically-designed surround microphones and calibrated mounting brackets can help provide a natural-sounding image and, at the same time, an interesting perspective.

Once you're able to utilize techniques that provide a realistic listening experience, there's no reason you can't break out of conformity to creatively shape and finely craft the emotional impact you want from your recordings.

Chapter 5 .... MICROPHONES

Downmixing

A consideration in surround, which is nearly inconsequential in stereo, is called downmixing. In the world of stereo we must always be aware that our mixes might be played back in mono—we adjust our settings so that our mixes sound good when the console's "MONO" button is pressed. The concern over mono compatibility is important, but far less so than it used to be in the past when television, radio, and PA systems were virtually all mono.

In the wonderful world of surround sound, mixes typically need to sound good in all surround formats as well as stereo and mono. Sometimes, we get to provide specifically crafted mixes in each format. The listener can select the mix they want to hear according to their playback format. Much of the time, our surround mixes are automatically downmixed from surround to stereo by a processor within the playback system. This is much more convenient and inexpensive; however, it can cause nightmares for the caring audio engineer. If each mix isn't cross-referenced through the automatic downmixing process, disappointment and anguish are likely.

LRFB

The LRFB (left, right, front, back) technique is a coincident setup because all of the mic capsules are positioned as close together as possible and they are positioned along the same axis. It uses two X - Y configurations—one pointing forward and one pointing backward.

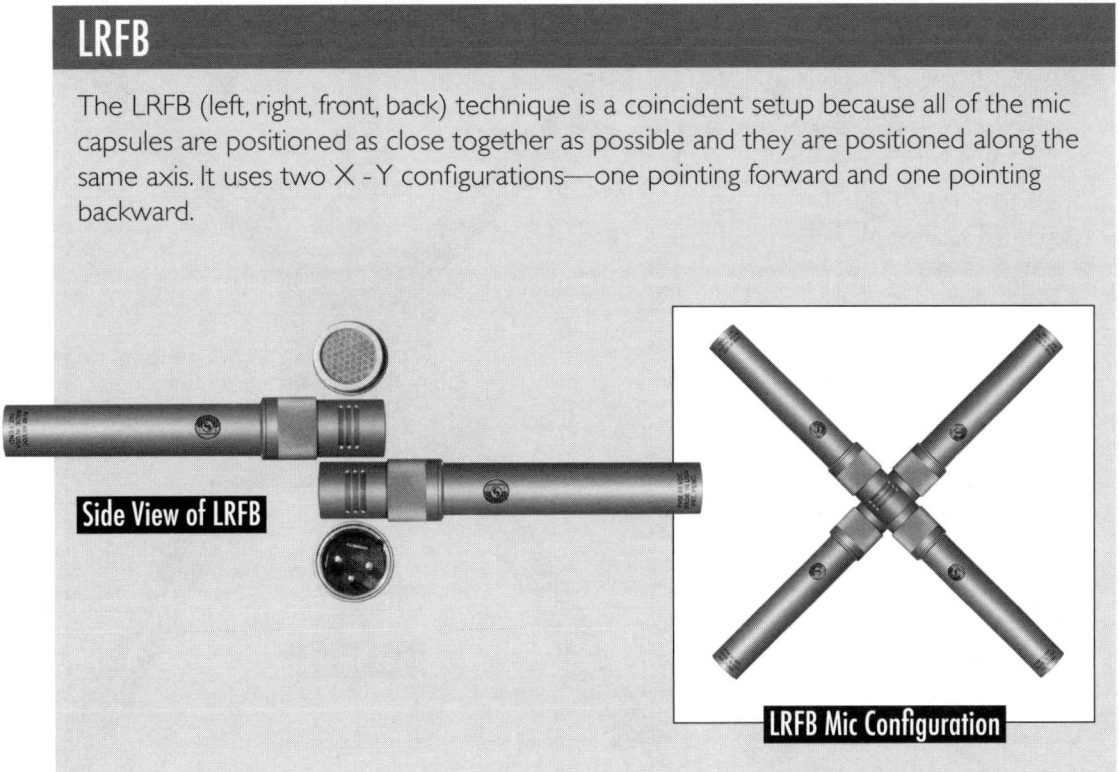

Side View of LRFB

LRFB Mic Configuration

The LRFB Technique

The LRFB (left, right, front, back) techniques utilizes two sets of mics in an X-Y configuration—one facing forward and one facing back. This setup downmixes very well because of its excellent phase coherence. This configuration adapts accurately to mono, stereo, quad, 5.1, and 7.1.

The VSA Tree

The use of commercially produced mic holders increases the reliability of surround mic techniques. The VSA tree is very adjustable and, over the course of time, the astute engineer can fine-tune the surround mic positions so that they provide a very natural sound, excellent phase coherence, and a quick and reliable setup.

Whether producing music or building audio scenes for film and television, surround sound is fun. In many ways it is easier to mix in surround because each audio ingredient can be quickly revealed in the 360-degree soundscape. However,

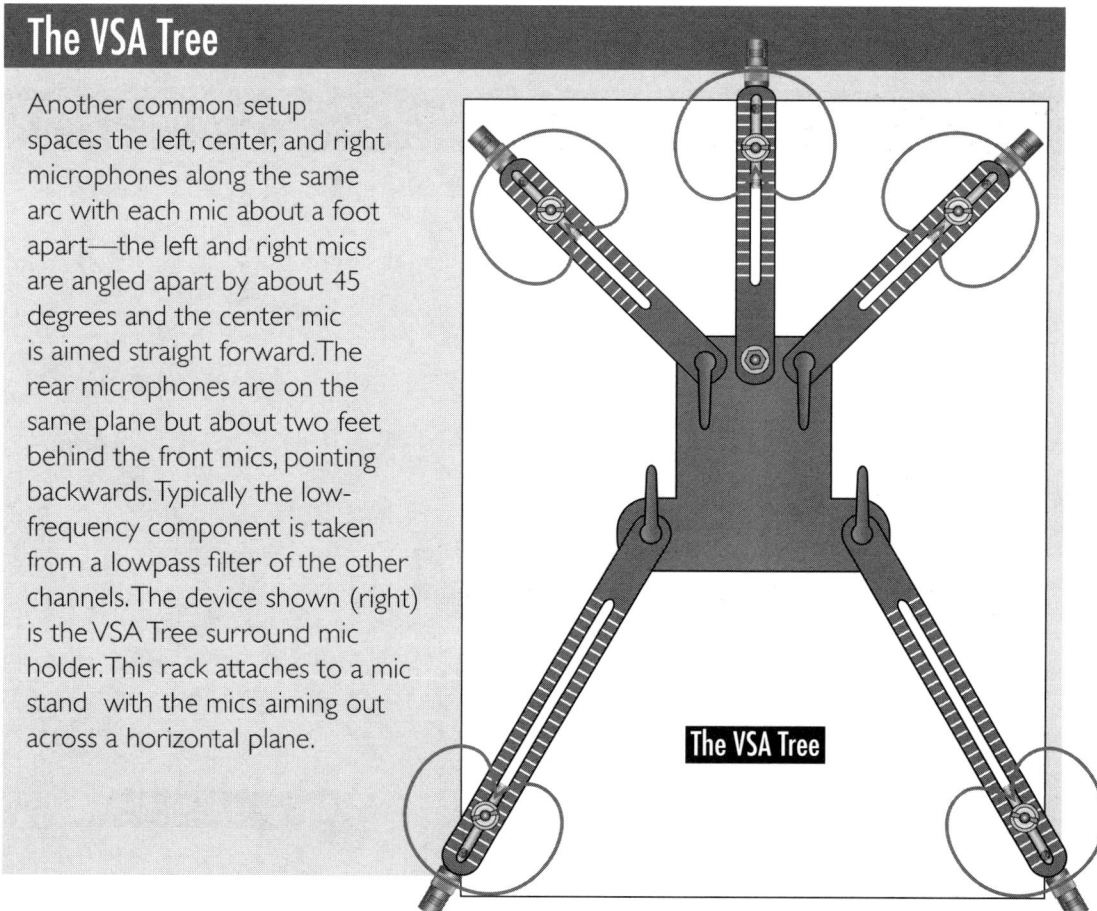

The VSA Tree

Another common setup spaces the left, center, and right microphones along the same arc with each mic about a foot apart—the left and right mics are angled apart by about 45 degrees and the center mic is aimed straight forward. The rear microphones are on the same plane but about two feet behind the front mics, pointing backwards. Typically the low-frequency component is taken from a lowpass filter of the other channels. The device shown (right) is the VSA Tree surround mic holder. This rack attaches to a mic stand with the mics aiming out across a horizontal plane.

The VSA Tree

Chapter 5 .... MICROPHONES

the techniques and insights that are required in a producing an excellent stereo mix are still very valuable in building an excellent surround mix.

CONCLUSION

You can cover most home recording situations if you have at least one good moving-coil mic and one good condenser mic. With these two options available, you can fairly consistently achieve professional-sounding results. Ribbon microphones have become increasingly important in the digital era—they can add depth and warmth to the somewhat clinical sound of digital audio.

This chapter was written with the intent to provide fundamental information that will immediately help you understand the practical applications of the three basic microphone types used in the studio.

Microphones & Mixers.. by BILL GIBSON

CHAPTER TEST

1. When selecting a microphone, it's most important to consider:
    a. directional characteristic
    b. operating principle
    c. response characteristic
    d. the sound of the mic for our recording

2. If a mic doesn't sound good on a particular source:
    a. apply equalization until it sounds the way you want
    b. apply compression
    c. try another mic
    d. add reverberation

3. A polar response graph indicates sensitivity in a _____scope and is interpreted as a _____ image.
    a. near-field, far-field
    b. graphic, mental
    c. 360-degree, circular
    d. 360-degree, three-dimensional

4. The term "on-axis" indicates that the source is anywhere along the axis that runs through the microphone from front to back.
    a. True
    b. False

5. A true cardioid pickup pattern is typically _____ than a hypercardioid or supercardioid pattern.
    a. wider
    b. narrower
    c. longer
    d. more round

6. When using a bidirectional microphone, sounds that are 90 degrees off-axis double in amplitude because they are able to be picked up by both sides of the capsule.
    a. True
    b. False

7. For most microphones, all frequency ranges tend to follow the same polar response pattern.
    a. True
    b. False

8. A _____ is any device that transforms one type of energy into another type of energy.
    a. transformer
    b. transducer
    c. pickup
    d. All of the above

9. Mics that use _____, whether moving-coil or ribbon, are dynamic microphones.
    a. magnetic induction
    b. carbon induction
    c. dynamic induction
    d. electronic induction

10. A _____ is a device that stores an electrical charge.
    a. resistor
    b. transistor
    c. capacitor
    d. condenser
    e. Both c and d

Chapter 5 .... MICROPHONES

225

11. In an SM57, the movement of the _____ around the magnet is what causes the changing flow of electrons that represents the _____.
   a. condenser, waveform
   b. copper coil, sound wave
   c. ribbon, audio waveform
   d. wire, pressure zone

12. Typically, the most fragile of the mic types is the:
   a. condenser
   b. moving-coil
   c. ribbon
   d. electret

13. Condenser mics require power to:
   a. amplify the capsule-level signal to mic-level
   b. power the amplifying circuitry in the mic
   c. provide polarizing voltage to the capsule plates
   d. All of the above

14. Most microphone capsules, free-floating in space, are _____. Their directionality is commonly created by _____.
   a. omnidirectional, ports in the microphone housing
   b. lightweight, phase anomalies
   c. omnidirectional, electrical polarization
   d. inductors, electrical polarization

15. In almost every circumstance, a mic with a flat frequency response curve will provide the best results.
   a. True
   b. False

16. In the modern recording world microphones are _____.
   a. usually condensers
   b. all high-quality so they all sound pretty good
   c. dynamic in their operating principle
   d. low-impedance devices

17. This mic technique uses two mics in a coincident physical relationship. Their capsules are as close together as possible, aimed across the same physical plane, and aimed at a 90-degree angle to each other.
   a. the Blumlein technique
   b. the MS technique
   c. the X-Y technique
   d. the spaced omni pair technique
   e. Both a and c

18. The _____ technique is the most flexible of the stereo miking configurations.
   a. X-Y
   b. mid-side
   c. spaced omni
   d. Blumlein

19. The LRFB technique is a _____ setup utilizing two _____ configurations—one pointing forward and one pointing backward.
   a. stereo, mono
   b. coincident, X - Y
   c. quadraphonic, arrayed
   d. near-field, pressure zone

20. An automatically generated stereo mix, which is derived from a 5.1 or 7.1 surround mix, is called a _____.
   a. downmix
   b. consolidated mix
   c. condensed mix
   d. compressed mix

Test answers are on page 293

Processors

It is fundamental to your recording success that you're completely familiar with the signal processing equipment. You need to recognize the sounds of these basic tools, and you need to know how to adjust their settings to fit each unique musical situation. It's surprisingly simple to learn the controls on most processors. Once you know how to use them, you possess knowledge that lets you operate similar units with minimal stress and maximum efficiency.

Each processor offers creative control and technical assistance. As we continue to build knowledge and skill, we'll see that each processor has many different and creative uses, and, try as we might, there are times when we find ourselves technically backed into a corner. In times like these, a thorough understanding of the right processor at the right time is very important.

According to Funk & Wagnall's Standard Dictionary, process, when used as a verb, means to treat or prepare by a special method. A signal processor is doing just that to our music—treating and preparing it in order to form an appealing, appropriate, and intelligible blend of textures.

My thesaurus shows that synonyms for process are filter and sift. These, too, give an accurate image of what signal processors do. If we can filter our music like we can filter light, we can start with one color and end up with another. In the music and recording industry, musical textures are often referred to as colors.

Microphones & Mixers .... by BILL GIBSON

Describing music and sounds verbally is a necessary skill. In the middle of a session, you will come up with some great ideas, and the more experienced you become, the more easily the ideas will flow. Your ideas are worthless if you can't verbalize them to the other musicians you're with. You don't need to use the most current jargon for a session to go well, but you must be sincere, proficient, and easy to get along with.

Notice that we consistently describe what we hear with terms normally used for things that we see, feel, or taste. Producers are notorious for using terms like dark, cold, bright, kickin', intense, sweet, etc. Describing the emotional impact of music involves describing far more than just what we hear. Good music is fundamentally a form of emotional expression and communicates to all feelings and senses. Be involved enough in your pursuits to walk the walk and talk the talk in a way that is sincere and easily understood.

SIGNAL PROCESSOR BASICS

A signal processor changes your musical signal for two basic reasons:
*   To enhance or craft an existing sound in a way that better supports the overall work.
*   To compensate for an inherent problem with a sound.
    This section covers the three main categories of signal processors:
*   Dynamic range processors
*   Equalizers
*   Effects processors (which include delays, reverberation, and multi-effects processors)

Communicating a Feeling

It's not always easy to communicate a subjective and artistic feeling with objective language. As you get better at producing great music, you'll need to be able to share your technical expectations effectively. The following list of descriptive terms, although common language, provide a starting point in the process of explaining your feelings in words that paint a visual picture on a sonic canvas.

As an example, you might recognize a signal that's been compressed with a 20:1 ratio, an extremely fast attack time of about 100 microseconds and a release time of 1.5 seconds, resulting in up to 20 dB of gain reduction, but until you

Chapter 6 ....PROCESSORS

can translate that to the word "squashed" you are out of the musical communication loop. As your technical understanding increases, you'll communicate in more technical terms; however, we capture art, which means we constantly deal with artists. They typically have no desire to clutter their minds with tech talk, which is why this following list of expressions is a springboard into building a vocabulary that tries to describe the sometimes indescribable:

Big: Containing a broad range of frequencies with ample clarity and sparkle in the highs and plenty of punch and thump in the lows. Usually contains large-sounding reverbs or large amounts of interesting reverb effects. Very impressive. Synonyms: huge, gigantic, large, monstrous.

Cool: The definition of cool changes with musical style. Very impressive, in a stylistically sophisticated way.

Dry: Without reverb or effect.

Edge: Upper frequencies of a sound that have a penetrating and potentially abrasive effect (typically 3 – 8 kHz). Used in moderation, these are the frequencies that add clarity and understandability.

Honk: See squawk.

Lush: Very smooth, pleasing texture. Often used in reference to strings that use wide voicings and interesting (although not extremely dissonant) harmonies. Typically includes a fair amount of reverb or concert hall.

Moo: Smooth, rich, and creamy lows.

Open: Uncompressed, natural, and clean with a wide dynamic range—a sound that can be heard through, or seen through, to use a visual analogy. In a musical arrangement, a situation where there is a lot of space (places in the arrangement where silence is a key factor). Each part is important and audible, and the acoustical sound of the hall can be appreciated.

Raunchy: Often slightly distorted (especially in reference to a guitar). A sound that doesn't include the very high frequencies or the very low frequencies. Earthy and bluesy. When referring to musical style, indicates a loose and simple but soul-wrenching performance.

Shimmer: Like sparkle in frequency content. Often includes a high-frequency reverberation or some other type of lengthened decay.

Sizzle: See sparkle. Can also include the airy-sounding highs.

Microphones & Mixers.. by BILL GIBSON

Sparkle: The upper frequencies of a sound. Includes the high bell-like sounds and upper cymbal frequencies from approximately 8 – 20 kHz. These are very high frequencies that add clarity and excitement.

Squawk: Midrange accentuation (approximately 1 kHz). Sounds a lot like a very small, cheap transistor radio.

Squashed: Heavily compressed. Put into a very narrow dynamic range.

Sweet: Similar to lush in that it is smooth and pleasing and includes a fair amount of reverb. Generally in a slightly higher register (above middle C). Pleasantly consonant.

Syrupy: Sweet, consonant sounds with ample reverberation. Often very musically and stylistically predictable.

Thump: Low frequencies. Especially, the lows that can be felt as well as heard (about 80 – 150 Hz).

Transparent: Nonintrusive. A sound that has a broad range of frequencies but doesn't cover all the other sounds around it. A sound that silence can be heard through.

Verb: Reverberation.

Wash: Lots of reverb that runs from one note to the next. This is common on string pads, where the reverb becomes an interesting part of the pad texture. A producer will often ask the engineer to bathe the strings in reverb, so the engineer gives the producer a wash of reverb.

Wet: Reverberation. Doesn't include the direct, original sound. To say something is very wet indicates that it's heard with a lot of reverb and not too much of the original, non-reverberated sound. Sometimes used in reference to other effects as well.

Similarity of Hardware and Software Controls

In our study of signal processors, for now, there is little distinction between hardware and software plug-ins. The controls are typically identical and the intended effect is the same. In fact, it's very common for a software signal processor to mimic the sound and functionality of a classic piece of hardware.

Sonically, hardware and software processors are capable of providing excellent results; however, if they're used inappropriately, either will suck the life out of your recordings. Knowledge is the key; practice is the routine. There's not necessarily a right and wrong way to use signal processors, but there is a common and

Chapter 6 ....PROCESSORS

231

uncommon way. I aim to help you understand these valuable tools so that you can operate them efficiently, or if you want to paint outside the lines, you should at least do it intentionally.

CONNECTING PROCESSORS

Whether you're using hardware, software, or software-based hardware, processors are typically connected to your system in one of only a few ways:

+ Channel inserts
+ Aux sends
+ Direct patches from instruments

Channel Insert

The channel insert on the mixer is an individual patch point on a channel. A piece of gear connected here becomes part of the signal from that point on. An

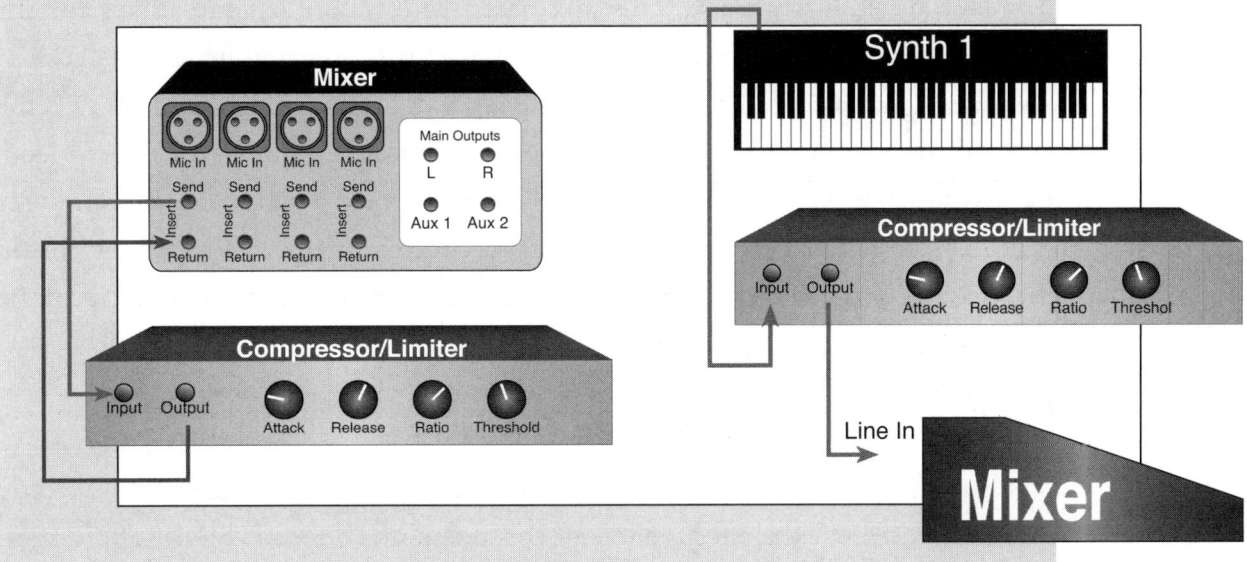

Patching the Dynamic Processor

Dynamic processors are typically inserted into a channel. The output of the processor, when patched into the channel return, supplies the dynamically altered signal back into the channel signal path—it's a permanent part of the sound from that point.

It's also common to patch a source directly into a dynamics processor before it reaches the mixer.

insert contains an output from the signal path, usually just after the input pre-amp, which is connected or routed to the processor input. Then, the output from the processor is connected or routed right back into the signal path at the return point of the insert. Although any device or plug-in with an input and output can be inserted into the signal path, this type of connection is most commonly used with dynamic processors.

Auxiliary Bus

An auxiliary bus, also called an *aux bus*, is nothing more than a separate mix from the available mixer channels that can be sent to a number of possible devices or locations. It has several valuable uses in the course of recording and mixing audio, including feeding various combinations of channels to effects devices, setting up sub groups, building alternate mixes for feeding different devices or monitor systems, feeding headphones, building mix stems for film and media use, and so on. One of its most frequent uses is feeding an effects device, such as a reverb or delay. Effects devices, though they can be inserted into individual channel signal paths, are most frequently connected to the output of an aux bus, often called an *effects bus*, so that varying levels of selected mix ingredients can be routed through the device or plug-in.

Once fed to the device input from the aux bus output, the routed signal is processed and then sent back to mixer via a dedicated effects return or simply through a regular mixer channel. Always try to keep the effects return separate from the original dry track. This aux-based connection scheme allows for the original dry track to remain intact and pristine, including any dynamics processing that might have been connected at the individual channel insert. However, the effected return can be blended into the mix for just the right balance and blend. In addition, plug-in effects processors place the highest demand on the computer processor, so rather than inserting several processor-intensive effects plug-ins on the mix, the user can use one or two bus-fed effects processor to cover most of the mix needs.

Plug the Instrument Directly into the Processor

It's also common, though not always efficient, to simply plug an instrument directly into the processor. The effectiveness of this technique depends on the application and the device. Many modern devices are designed to accept several different line-

Chapter 6 ....PROCESSORS

and instrument-level signals. The viability of routing the instrument through the processor in real-time as the musical performance is being recorded is very dependent on the computer, the plug-in, and the size of the project file. Computer-based systems often route the audio through the software recorder before the performer can hear it playback. This process can take a few milliseconds to achieve and there is frequently an audible delay from the onset of the performed note to the return of the playback of that note to the performer. This delay is referred to as *latency*. Modern DAW software-based recording systems are capable of extremely low latency; however, careful parameter adjustments are typically required to minimize or negate latency problems.

DYNAMICS PROCESSOR OPERATION

In general, dynamic processors are used early in the recording process and at the beginning of the signal path, whereas effects processors are best reserved for mixdown. Since dynamics processors control dynamic range, this is an opportune time for a definition. Dynamic range is the distance in decibels from the softest sound to the loudest sound. If an orchestra plays its loudest note at 115 dB and its softest note at 20 dB, its dynamic range is 95 dB—the loudest sound (115 dB) minus the softest sound (20 dB).

Dynamic Range

Dynamic range processors are often subtle in the effect they have on a musical sound, and in most situations, the listener shouldn't be aware that anything out of the ordinary is going on.

These processors (compressors, limiters, gates, and expanders) all work in a very similar way and—whether hardware or software—have very similar, if not identical, controls from unit to unit. The task for any dynamic processor is to change the distance, in volume, from the softest sound to the loudest sound, or to alter the dynamic range.

The VCA

The central operator in most of the analog dynamics processors is the voltage-controlled amplifier (VCA). Its name is almost its definition. Inside each processor is an amplifying circuit that turns the volume up and down as it senses more

Microphones & Mixers.. by BILL GIBSON

or less voltage—it's a voltage-controlled amplifier. The changing levels in your musical signal determine the amount of voltage sent to the input of the VCA. Whether you're using a classic analog dynamics processor, a fully digital plug-in, or a hybrid digitally-controlled analog device, the concept of the VCA provides an excellent point of reference and a pathway to understanding.

The VCA is capable of responding to increases or decreases in input voltage by increasing or decreasing the output of the amplifying circuit. How it responds depends entirely on its intended function and the user-adjustable parameters. Most dynamic range processors set the VCA so that, in relation to unity gain, it turns the signal level down in response to a specific voltage change, then turns it back up again according to the parameter settings. However, there are a few applications where the VCA actually boosts the signal above its unity gain setting.

Keep in mind that software plug-ins, though they might digitally mimic the action of the VCA, still respond to level changes according to the user-adjusted settings in the same manner as hardware processors.

Dynamic range processors are typically patched into the signal path of the microphone, instrument, or recorder track via the channel insert (on the mixer or the patch bay). They're also commonly patched inline between the source and the mixer.

Substitutes for the VCA

The VCA (Voltage-Controlled Amplifier) is an analog amplifier, which is controlled by variations in voltage. This is the primary dynamic control circuit in most compressors, limiters, gates, and expanders.

There are options to the VCA—they include:

- The DCA (Digitally-Controlled Amplifier) is an analog amplifier, which is controlled by variations in digital data. Some of the most highly regarded consoles combine the warmth and purity of analog circuitry with the precision and flexibility of digital control. In addition to precision, digital control data is easily stored, automated, and recalled.
- The Optical Level Control offers a very smooth and precise level control system utilizing a light-dependent resistor called an opto-isolator. When light shines on this special resistor, some of the signal is shunted to ground, which reduces the level. Processor parameters are dependent on reactions to light intensity.

Chapter 6 ....PROCESSORS

♦   Data Control operates in the digital domain to control dynamics through mathematical calculations. The actions of the VCA or DCA are simulated according to digitally encoded instructions. As long as the algorithms are well founded, digital manipulation is very accurate and efficient.

Although there are multiple possible devices that control the dynamic processor level, we'll typically refer to the VCA generically as the controller.

COMPRESSOR/LIMITER

The compressor is an automatic volume control that turns loud parts of the musical signal down. When the VCA senses the signal exceeding a certain level, it acts on that signal and turns it down.

Imagine yourself listening to the mix, and every time the vocal track starts to get too loud and read too hot on the meter, you turn the fader down and then back up again for the rest of the track. That is exactly how a compressor works.

A compressor is a useful tool when recording instruments with a wide dynamic range. Compressors are typically used on vocals, bass, or any other instruments with a wide dynamic range.

Again, the VCA in a compressor only turns down in response to a signal and then turns back up again; it doesn't turn up beyond unity gain (the original level).

Why Do We Need a Compressor?

We need a compressor to protect against overly loud sounds that can overdrive electronic circuitry, oversaturate magnetic tape, or overdrive digital or analog inputs. A compressor also helps even out the different ranges of an instrument. Instruments like brass, strings, vocals, and guitars can have substantially different volumes and impact in different pitch ranges. These ranges can disappear, then suddenly jump out in a mix. A highly skilled and very focused engineer might catch many of these variations in level, but a compressor is often more reliable and less intrusive. A compressor can also even out volume differences created by an artist changing their distance from the mic.

The Resulting Effect of Compression

Since we've put a lid on the loud passages, we can therefore print the entire track, with a stronger signal, to the recorder. We are able to move the overall signal into

Microphones & Mixers.. by BILL GIBSON

a tighter dynamic range, which is especially useful in a commercial popular genre, which is typically played on radio or television. This provides a better signal-to-noise ratio and, when used properly, contributes to the achievement of a professional-sounding mix.

When used correctly, compression doesn't detract from the life of the original sound. In fact, it can be the one tool that helps that life and depth to be heard and understood in a mix. Imagine a vocal track. Singers perform many nuances and licks that define their individual style. Within the same second, they may jump from a subtle, emotional phrase to a screaming-loud, needle-pegging, engineer-torturing high note. Even the best of us aren't fast enough to catch all of these changes by simply riding the input fader. In this situation, a compressor is needed to protect against excessive levels.

This automatic level control gives us a very important by-product. As the loudest parts of the track are turned down, we're able to bring the overall level of the track up. In effect, this brings the softer sounds up in relation to the louder sounds. The subtle nuance becomes more noticeable in a mix, so the individuality and style of the artist is more easily recognized, plus the understandability and audibility of the lyrics are greatly increased.

.... Video Example 6-1

Demonstration of Compression and Makeup Gain

Analog Versus Digital Application

If you're using analog tape, compression can be important during tracking. Since the noise floor is high when using magnetic tape, an acceptable way to keep your recordings free from tape noise is to compress the tracks into a tighter dynamic range and record the hottest level possible to tape. In effect, this brings the softer sounds up farther away from the noise floor, so if they need to be turned up in the mix, they're as clean and noise-free as possible. Modern tape formulations are capable of recording very hot signals, which helps keep the recorded audio farther from the noise floor than when recording to more vintage formulations. Because of this, compression can easily be used more sparingly today than in the early days of analog recording. As an aside, many fantastic and powerful recordings have been made throughout history with little or no use of compression. The way compression is implemented in the recording process is dependent on creative and

The End Result of Compression

Ideally, the end result of compression is that the loudest portions of the signal sound about the same as normal, but the softest portions seem louder.

A compressor automatically turns the loud parts down as soon as it senses their levels. Once the signal passes the user-set threshold, the VCA acts on the signal according to the ratio setting.

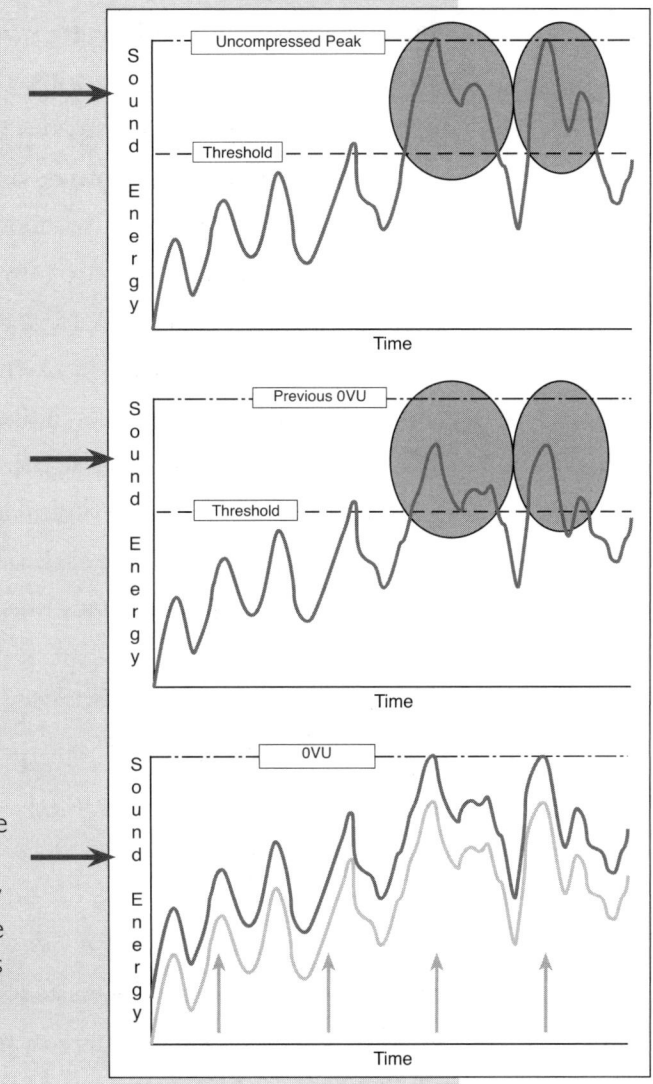

This graph represents the compression of the wave above. Everything below the threshold is unaffected. Everything above the threshold is reduced in level according to the ratio setting.

The pink line represents the level of the compressed signal above. Once the compressor has turned the loudest part of the track down, the entire track can be turned up so the overall level still reaches 0 VU. The red line represents the new level. Notice that the softest parts of the track are louder (consequently easier to hear in the mix) as the entire level increases.

musical considerations and the resulting technical decisions made by the engineer and producer. It is usually best to use only the amount of compression required to achieve a desired artistic result. Over-compressing a recording can quickly degrade the audio quality and reduce the overall power and impact of the final mix.

Applying compression during mixdown, when using an analog multitrack, increases the likelihood that the noise level will audibly raise and lower during the mix. As the compressor rides the level of important tracks, like the lead or

Microphones & Mixers.. by BILL GIBSON

backing vocals, the airy hiss (called tape noise) tends to turn up during the softer, more open portion, and then it disappears during the loud and complex sections. When the VCA turns the tape track back up during the soft passages, the tape noise is audibly increased, too. We hear this noise turning up and down as the signal crosses the threshold, and the VCA reacts by turning up and down. This is one of the adverse effects of compression. The sound of the noise turning up and down is called *pumping* or *breathing*.

In the digital realm, noise is not really the issue, but there are still a few good reasons to use compression when tracking or mixing.

- In a commercial mix, each ingredient is carefully placed, so, whether digital or analog, compression can help define the position and focus of a track.
- Whether a vocal or instrumental track, compression helps increase the relative level of the nuance, resulting in increased understandability and impact.
- In the digital realm bit depth equates to amplitude resolution, so it's important to record near full digital levels at some point to ensure maximum definition. Compression helps us to record at higher digital levels with less concern about exceeding maximum level and achieving clipping or digital distortion.

To Compress or Not to Compress—That Is the Question

Some engineers use a lot of compression and limiting; other engineers don't use any. In commercial popular music, the use (and overuse) of dynamics processors is common. Most engineers want their mixes to sound the loudest, in relation to other commercial mixes—it becomes an obsession for most. Therefore, each ingredient is tightly placed in its own dynamic, pan, and frequency range. Then the entire mix is limited so it's as loud as technically possible.

The only problem with compression obsession is that, in an effort to create a mix that's big and loud, the result can be narrow and lifeless, especially when there are a lot of mix ingredients. Good production, great music, and excellent recording technique are the real keys to a big sound.

Many producers and engineers don't like the sound of the compressor working. They prefer the natural space around a pure sound, and they rarely incorporate compression in tracking or mixdown. However, though they don't use automatic dynamic processors, they typically ride the levels of the tracks as they're being recorded or mixed—in effect, they're acting as human, real-time compressors.

Controls on the Compressor/Limiter

Almost all compressor/limiters contain the same control options, whether hardware or software. Once you understand the functions on one compressor/limiter, you'll find seamless transition to another. The unit pictured here contains the basic controls: attack time, release time, threshold, ratio, output level, peak/RMS, knee, and meter function.

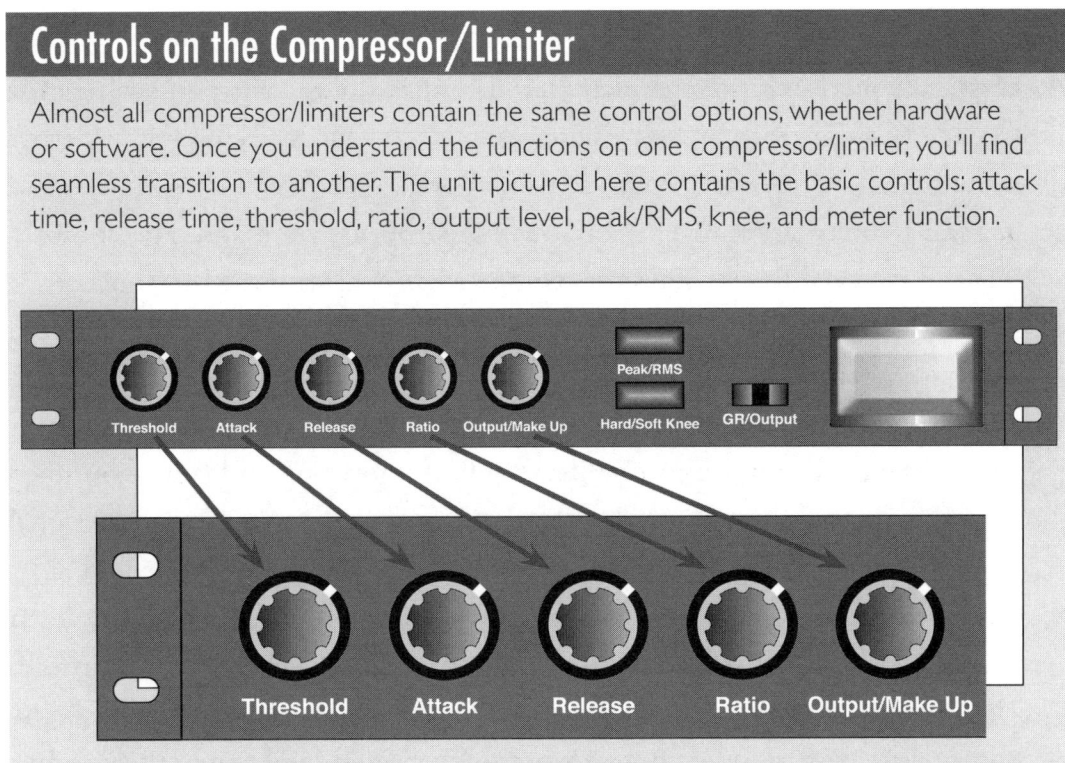

It's common for engineers to use dynamics processors very conservatively during tracking, adjusting compression for minimal gain reduction, then assessing the need for further compression during mixdown. This technique is much more realistic in the digital realm.

You need to assess the viability of using compression for each instance during the recording process. Evaluate the music, the intended audience, the musicians, the genre, and your own personal taste. There are a lot of cases where I, personally, like the sound of a track or song that's been compressed. Other times I like a more natural sound. Also, just because you're using a compressor, doesn't mean you have to overuse it.

COMPRESSION PARAMETERS

There are five controls common to most compressors: threshold, attack time, release time, ratio, and output level.

Once you see how these work, you can operate any compressor, anywhere, anytime. To make it even better, these controls are easy to understand, and they do just what they say they do.

Microphones & Mixers.. by BILL GIBSON

Threshold

As amplitude increases, voltage increases. The threshold is the point where the compressor begins to recognize the signal amplitude. Once the compressor recognizes the signal—when the amplitude rises above a certain voltage—it begins to act in a way that is determined by the attack time, release time, and ratio controls.

There are two different ways that compressors deal with the threshold:

+ One way boosts the signal up into the threshold. Picture yourself in a room with an opening in the ceiling directly overhead. You represent the signal, with your head being the loudest sounds. The opening represents the threshold of the compressor. Imagine that the floor moves up and you begin to go through the opening. That's the way that some compressors move the signal into the threshold—they turn it up until it goes through the threshold.

+ The other way compressors deal with the threshold is by moving it down into the signal. Picture yourself in a room with an opening directly overhead. Now the ceiling moves down until you're through the opening. This is the other way the threshold control works—the signal level stays the same but the threshold moves down into the peaks.

No matter which way the threshold works, it's the part of the signal that exceeds the threshold that's processed. Once the signal is through the threshold, the VCA turns down just the part of the signal that's gone through, leaving the rest of the signal unaffected. The portion that's above the threshold will be turned down according to how you have set the remaining controls (attack time, release time, and ratio).

The Moving Threshold

The threshold control moves the threshold up or down in relation to the energy of the signal. Anything above the threshold is acted on by the VCA. Anything below the threshold is left unaffected.

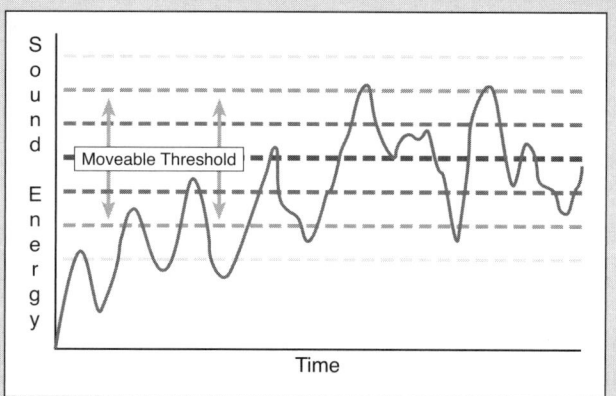

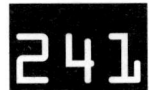

Attack Time

The attack time controls the amount of time it takes the compressor to turn the signal down, once it's passed the threshold. If the attack time is too fast, the compressor will turn down the transients. This can cause an instrument to lose life and clarity. On a vocal, for instance, if the attack time is too fast, all of the sibilant sound, such as "T," "Ch," and "S" will start to disappear. On the other hand, if the attack time is too slow and the vocal is very compressed, the sibilant sounds will fly through before the VCA is able to compress them and, therefore, sound exaggerated.

Attack Time, Understandability, and Punch

The attack setting provides a means to adjust the relative level of the initial portion of an audio source. In a vocal passage, the initial transient sounds—especially the sounds "s," "t," and "k"—offer two possible complications for recording:

1. If the vocalist has a natural abundance of sibilance, the recorded track might take on a harsh character. These transient sounds, called sibilance, can cause distortion of analog tape and irritating effects when reverberated, and they can even overdrive electronic circuitry. In this case, a fast attack time, during compression helps smooth out the sound—the track will settle into the mix better.

2. If the instrumental bed is very percussive, and if the vocal sound contains understated sibilance, the lyrics might be lost in the mix because they're not understandable. In this case, try compressing the vocal track using a slower attack time; the compressor will let the sibilance pass through unaltered, yet the rest of the word will be compressed according to the control settings.

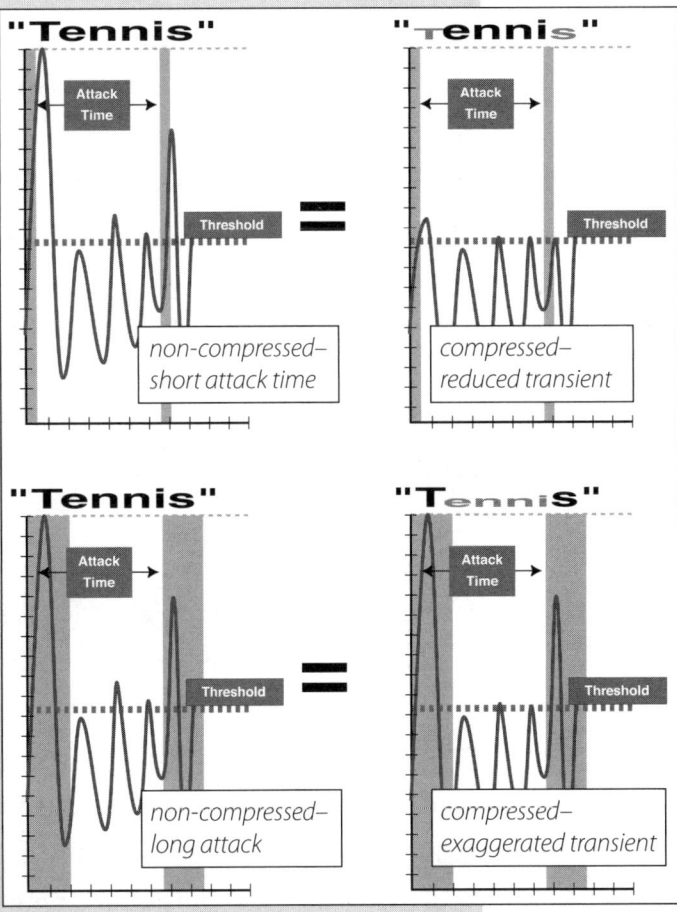

Microphones & Mixers.. by BILL GIBSON

······························ Audio Example 6-1

Ss and Ts

······························ Audio Example 6-2

Exaggerated Ss and Ts

Variations in the attack time setting help diminish or accentuate the relative attack of instruments like guitar, bass, piano, or drums. Long attack times adjust average levels; short attack times adjust peak levels.

Specific attack time limitations vary between processors, though they typically range from 0.1 ms to 200 ms. One characteristic of an expensive compressor is fast attack time capability. Also, some compressors have the attack time fixed for a specific purpose, like vocals.

Release Time

Release time is the time that it takes for the compressor to let go, or turn the signal back up, once it's below the threshold. The release time might be as fast as 50 ms or as slow five seconds.

Fast release times work well with fast attack times to control peak levels. Slow release times work well with slow attack times to control average levels. There is no practical value to adjusting the attack time so it's slower than the release time.

Long release times with severe compression can result in increased sustain. With the proper setting of the threshold, release, and attack time, a guitar, for example, can benefit by increased sustain. Over time, as the VCA turns the signal back up to its original level, an otherwise quickly decaying signal maintains its sustain longer.

For a natural and unobtrusive sound, set the attack time relatively fast and the release time relatively slow. Each instrument or voice is different, so there's still importance placed on listening while you adjust these controls.

Ratio

Once the compressor starts acting on the signal, the ratio control determines how extreme the VCA action will be. The ratio is simply a comparison between the level that goes through the threshold and the output of the VCA; it's expressed as a mathematical ratio (10:1, 3:1, etc.). The first number in the ratio indicates how

Compression with a 3:1 Ratio

The threshold in the top graph is set so that the peak sound energy level exceeds the threshold by 12 dB. The VCA turns the signal (above the threshold) down according to the ratio. With the ratio set at 3:1, the VCA only allows 1 dB of increase for every 3 dB that exceed the threshold. The original signal exceeded the threshold by 12 dB (with no compression), but the compressor only allows a 4-dB peak when the ratio is set at 3:1 (bottom graph).

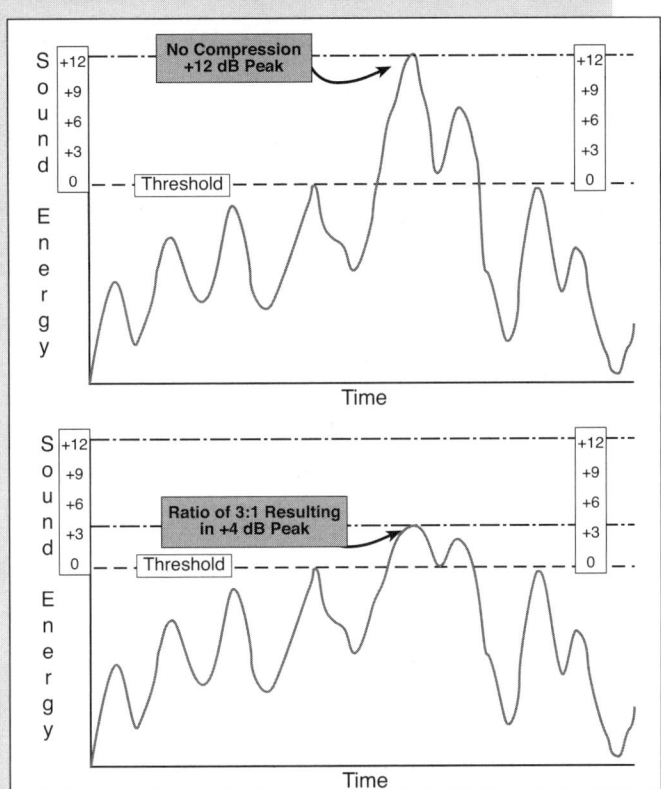

many dB of input increase will result in 1 dB of output increase. The higher the ratio, the greater the compression.

If the threshold is adjusted so that the loudest note of the song exceeds the threshold by 3 dB, and the ratio is 3:1, the 3-dB peak is reduced to a 1-dB peak—the gain is reduced by 2 dB. Using that same 3:1 ratio, if you input a 12-dB peak, the unit would output a 4 dB peak—still a ratio of 3:1, and the gain is reduced by 8 dB.

Output Level

The output level control, also called *makeup gain*, makes up for reduction in gain caused by the VCA. If the gain has been reduced by 6 dB, for example, the output level control is used to boost the signal back up to its original level.

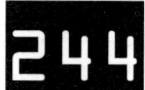

Microphones & Mixers.. by BILL GIBSON

The Difference between a Compressor and a Limiter

It's the ratio setting that determines the difference between a compressor and a limiter. Ratio settings below 10:1 result in compression. Ratio settings of 10:1 and above result in limiting. That explains why most manufacturers offer combined compressor/limiters. Extreme compression becomes limiting.

HARD KNEE VERSUS SOFT KNEE COMPRESSION/LIMITING

Hard knee/soft knee selection determines how the compressor reacts to the signal once it passes this threshold and the amplifier circuitry engages. Whereas the ratio control determines the severity of compression, the knee determines how severely and immediately the compressor acts on that signal.

When the compressor is set on soft knee and the signal exceeds the threshold, the amplitude is gradually reduced throughout the first 5 dB or so of gain reduction. When the compressor is set on hard knee and the signal exceeds the threshold, it is rapidly and severely reduced in amplitude. The hard knee/soft knee settings are still dependent on the ratio, attack, release, and threshold settings. The knee setting specifically relates to how the amplifier circuitry reacts at the onset of compression or limiting.

The difference between hard knee and soft knee compression is more apparent at extreme compression ratios and gain reduction. Soft knee compression is most useful during high-ratio compression or limiting. The gentle approach of the soft knee setting is least obvious as the compressor begins gain reduction. Hard knee settings are very efficient when extreme and immediate limiting is called for, especially when used on audio containing an abundance of transient peaks.

Typically, soft knee compression is more gentle and less audible than hard knee. Try this setting on a lead vocal or lyrical instrument for inconspicuous level control.

Hard knee dynamic control is more extreme and much less sonically forgiving. Try hard knee limiting when absolute level control is necessary.

Chapter 6 ....PROCESSORS

245

Hard Knee Versus Soft Knee Compression/Limiting

Once the compressor senses signal above the threshold and the attack time has passed, the level-control circuitry begins to respond. A hard knee setting activates the dynamic process immediately; a soft knee setting gradually engages dynamic control during the first 5 dB or so. Typically, soft knee compression is more gentle and less audible than hard knee —try this setting on a lead vocal or lyrical instrument for inconspicuous level control.

Hard knee dynamic control is more extreme and much less sonically forgiving. Try hard knee limiting when absolute level control is necessary.

The dynamic action matches the word picture—soft knee creates a gently rounded level adjustment; hard knee creates a sharp angle.

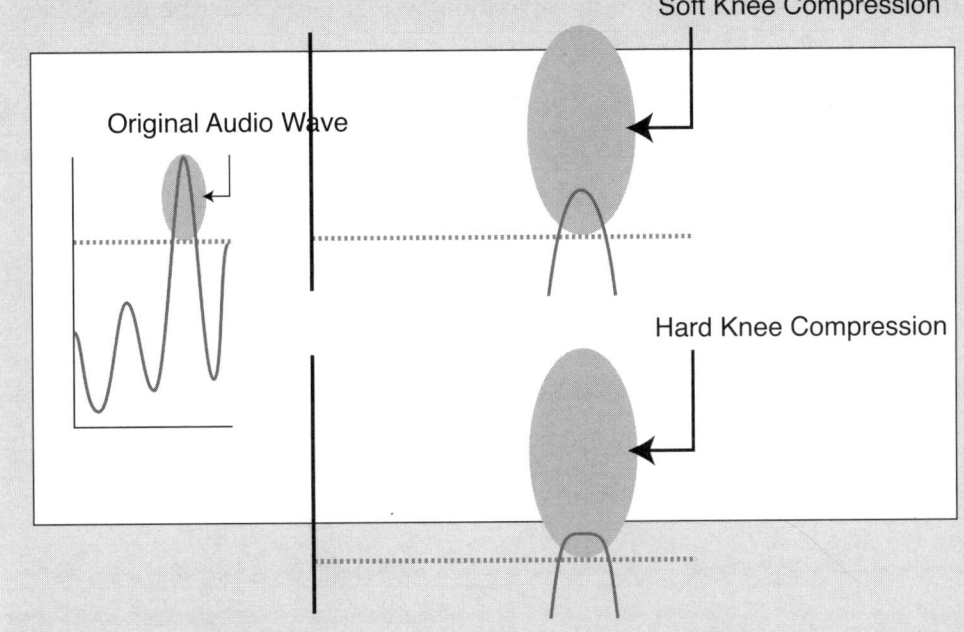

Soft Knee Compression

Original Audio Wave

Hard Knee Compression

PEAK/RMS DETECTION

RMS refers to average signal amplitude, based on the mathematical function of the Root Mean Square. Peak refers to immediate and transient amplitude levels, which occur frequently throughout most audio recordings. The Peak/RMS setting determines whether the compressor/limiter responds to average amplitude changes or peak amplitude changes. RMS compression is more gentle and unobtrusive than peak compression. Peak compression is well suited to limiting applications. It responds quickly and efficiently to incoming amplitude changes containing transient information.

246

Microphones & Mixers... by BILL GIBSON

SIDE CHAIN

The side chain provides an avenue for activating the level-control circuitry from a source other than the audio signal running the unit.

Any audio source can be patched into the side chain input for creative applications. However, it's common to run a split from the audio signal through an equalizer, then back into the side chain. In this way, the equalizer can be boosted at a specific frequency and cut at others, allowing the user to select a problem frequency to trigger gain reduction. This technique works very well when low-frequency pops or thumps must be compressed while the rest of the audio signal is left unaffected, or when certain high-frequency transients must be controlled.

Listen to the following audio examples highlighting the sonic impact of different compressor/limiter settings. The acoustic guitar is often compressed, and in these examples it provides an excellent comparison. With its clean, clear sound and transient attack, the parameter adjustments are very apparent.

· Audio Example 6-3

Acoustic Guitar - No Compression

This acoustic guitar, recorded without compression, has clean sound; however, it has a wide dynamic range. Notice the difference between the level of the loudest sound and the softest sound.

· Audio Example 6-4

Acoustic Guitar - Variations in Attack Time Settings

Notice the change in the attack of each note. By increasing and decreasing the attack time, intimacy and sonic impact change dramatically.

· Audio Example 6-5

Acoustic Guitar - Long and Short Release Times

With the release time set too short, the processor is continually active, risking sonic degradation. With the release time set too short, the level changes become very noticeable.

· Audio Example 6-6

Acoustic Guitar - Ratios: From Compression to Limiting

Chapter 6 ....PROCESSORS

247

Side Chain Control of the Level-Changing Circuitry

Use the side chain send and return to control dynamics from an external source. This illustration demonstrates a common use for side chain inserts during compression or limiting.

Notice the analog inputs and outputs are connected in the normal manner—the actual audio signal does not pass through the equalizer. The side chain send routes the audio to the equalizer, then the signal is equalized to accentuate a problem frequency.

Once it has been equalized, the signal is patched back into the unit through the side chain return. When the side chain circuit is enabled, the compressor's level detection circuit reacts to the equalized signal instead of the regular analog input.

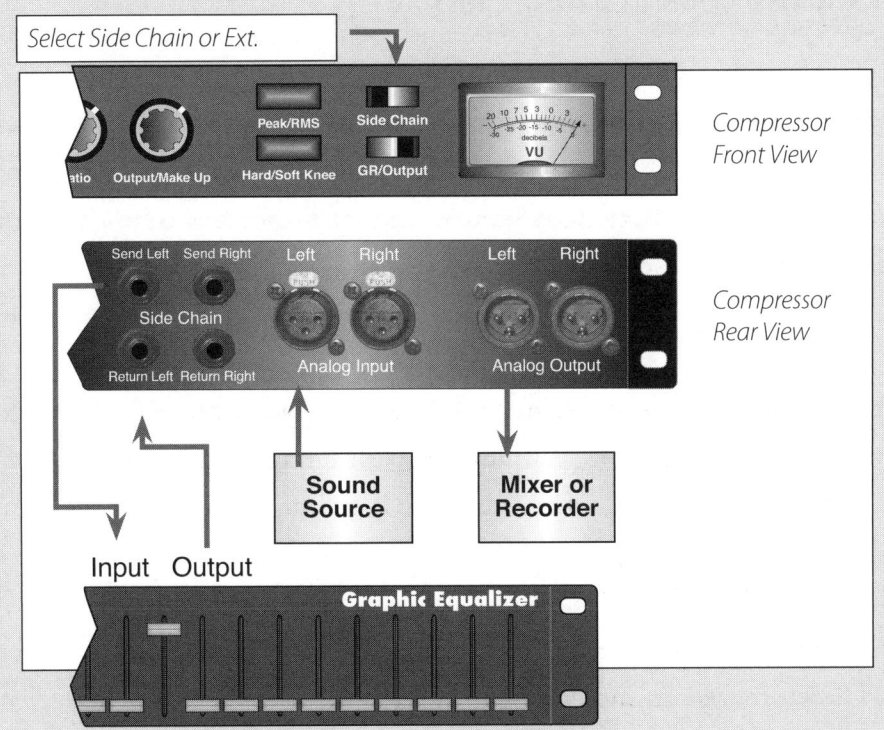

Each ratio setting provides a different result, from gentle gain control to the brick wall.

• Audio Example 6-7

Acoustic Guitar – Adjusting the Threshold for Optimum Sonics

Listen to the changing sound as the threshold moves down into the signal amplitude. If the threshold is too high, there is no dynamic compression. If the threshold includes too much of the amplitude, the level-changing circuitry (VCA,

DCA, Optical Amp, etc.) is always working; this typically produces a thin, weak, or strained sound.

. . . . **Audio Example 6-8**
Acoustic Guitar – Adjusting the Knee and Peak/RMS Settings

With the ratio set at 7:1, attack time at 10 ms, release time at .5 seconds, and the threshold set for 6 dB of gain reduction, notice the sonic difference as I switch from soft to hard knee, and from peak to RMS detection.

METERS ON THE COMPRESSOR/LIMITER

Compressor/limiters utilize various systems for metering gain reduction, input levels, and output levels. Some devices offer separate meters for each function, whereas several units utilize a multipurpose meter that switches between functions. Either system is functionally simple, and it's important to use these meters to help ensure optimal use of the device.

Input Level Meter

Ideally, the input level meter verifies the proper signal strength as it enters the device; however, many compressor/limiters don't have one. Since compressor/limiters are usually patched inline directly or through an insert, you can take advantage of the meters on your mixer.

When the compressor limiter is in bypass mode (most units have a bypass switch) the input typically defaults to unity gain with the output (no boost and no cut). Therefore, as you increase gain reduction, you should be able to simply boost the output level to maintain the original level or, if your device doesn't have an output level control, you can usually make the gain up by increasing the channel gain trim.

Output Level Meter

The output level meter is simply fed by the output level control. Use it to verify that the signal level is correct at the output of the device. It's common to set the output level so that it matches the input level—both at unity gain.

VU Meter Versus LEDs Versus Onscreen

Each dynamic processor provides a method to measure the amount of gain reduction occurring at any given time. Whereas, a typical meter reads from left to right to indicate the amount of signal present, a compressor/limiter meter typically moves from right to left to indicate the amount of signal decrease (in dB).

When a traditional VU meter indicates gain reduction, there is no level change as long as the meter is resting at the far right side. As the level is decreased, the meter moves to the left—the numbers on the meter represent decibels of gain reduction.

When a series of LEDs are used to indicate gain reduction, each LED that illuminates indicates more gain reduction. The numbers under each LED show the amount of gain reduction.

Computer-based compressor/limiters use an onscreen version of either of these metering systems.

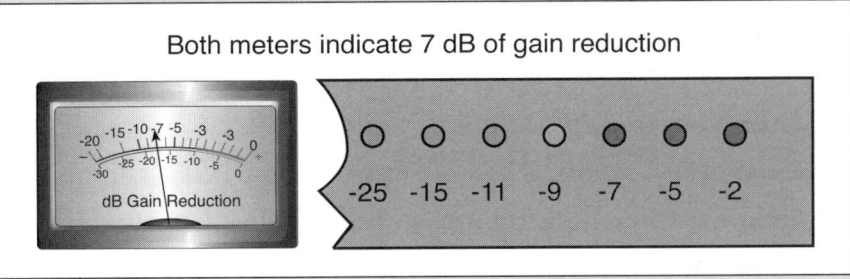

Both meters indicate 7 dB of gain reduction

Gain Reduction Meter

Gain reduction refers to the amount that the VCA has turned the signal down once it crosses the threshold.

To meter gain reduction, some compressor/limiters use a series of LEDs and others use a VU meter. Typically, LEDs light up from right to left, indicating how far the unit has turned the signal down. Each LED represents two or more dB of gain reduction.

If your compressor has a VU meter, 0 VU is the normal (rest) position on a meter used to indicate no gain reduction. As the compressor turns down, the needle moves backwards from 0 to indicate the amount of gain reduction. A –5 reading on the VU indicates 5 dB of gain reduction.

·· Video Example 6-2

Same Source through Different Compressor/Limiter Settings

Microphones & Mixers... by BILL GIBSON

THE FUNCTION OF A COMPRESSOR

The goal of the compressor is to gently ride the signal level in the same way that a human engineer would ride the fader while listening to the track playback. In most cases, the listener shouldn't even realize there's a processor being used.

Technical Effect

Technically speaking, a compressor allows the engineer to record an entire track at a hotter level than if the compressor were not included. If the compressor decreases the level of the hottest part of the track by a 6 dB, the entire track can be recorded 6 dB hotter—making up the reduced gain—without over modulating, saturating the tape, or exceeding the maximum digital recording level. Since the compressor should be transparent and seamless as it controls the maximum output level, the result is a more visible, audible, and apparent audio track, especially during the passages containing the least amplitude. In other words, the loud passages should exhibit minimal sonic effect, while the soft passages should be louder than they'd have been without the compressor. Though the compressor/limiter actually controls the loudest passages, the net result for the listener is an increase in the level of the softer passages.

Listen to this example of a vocal phrase: first, the original non-compressed recording, then the same phrase compressed by 6 dB.

·· Audio Example 6-9

Vocal Compressed by 6 dB

Musical Effect

Musically speaking, the compressor is very useful. The lead vocal track, for example, should be audible, understandable, and apparent throughout most popular commercial songs; this is precisely the result of compression. A compressed lead vocal track typically sounds more up front in the mix; it remains within its intended dynamic range.

Bass guitar is nearly always compressed. The low-frequency range of the bass contains an abundance of energy; therefore, it produces an abundance of amplitude. Left unchecked, this abundance of low-frequency energy can dominate the overall mix level. For example, when the bass part is particularly strong the overall

Chapter 6 ....PROCESSORS

251

mix level might be artificially hot. When the bass is properly compressed, however, that energy is kept in check, the bass remains consistently supportive of the mix, and the mix level can be increased.

Any instrument with a wide dynamic range can benefit from compression. However, certain classical and orchestral recordists use little or no compression. The natural dynamic range of the orchestra, piano, symphony, or instrumentalist adds emotional impact and reality to the recording. Many of these recordings are listened to in an environment conducive to such dynamic range: a living room, family room, media room, or listening space designed specifically for the enjoyment of music. In these instances, the natural dynamic range can be appreciated. On the other hand, many commercial popular recordings are listened to in a car, grocery store, mall, or other high ambient noise environment. Any dynamic subtleties might be lost when listened to in these places, so it's important that all the mix ingredients reside within the audible audio spectrum. Therefore, compression is very appropriate for these recordings.

THE FUNCTION OF A LIMITER

A limiter and compressor perform the same basic task, although a compressor typically controls level and amplitude in a soft and gentle manner, while a limiter controls level and amplitude in an extreme way. The choice to use compression or limiting is purely a musical one. For example, when recording a bass guitarist with a very consistent playing style, there might be little need for compression. However, if the same bassist slaps or snaps during a particular take, a limiter could help level out the amplitude peaks caused by these aggressive musical instances.

Limiters are often used to control the level of an entire mix. An excellent mix typically contains several transient peaks (levels that exceed the average level of the entire mix). Although the limiter ignores the majority of the program material (audio that doesn't exceed the threshold), a peak which exceeds the threshold will be turned down quickly. Through the use of limiters, most commercial recordings maintain a constant and aggressive level and amplitude. A master mix might peak at 0 VU; while the limited mix also peeks at 0 VU, the only difference is that the limited mix sounds louder. A good limiter operates in a way that is imperceptible to most listeners. It reacts quickly to transient peaks and maintains a full, impressive, aggressive sound throughout the limiting process.

252

Microphones & Mixers.. by BILL GIBSON

..............................Video Example 6-3

Meters Compared during Limiting

This video segment demonstrates the limiting of a complete mix. The meter on the left shows the output level of the non-limited audio. The meter on the right shows the output of the same audio, limited by 6 dB.

Now listen to the difference in volume between the two mixes, once the limited mix is boosted 6 dB to make up the gain reduction.

..............................Audio Example 6-10

Making Up Gain That's Been Reduced

PROPER USE OF THE COMPRESSOR/LIMITER

Typically, the threshold is set above normal operating levels—most of the time, there should be no gain reduction. If there's always gain reduction, the VCA is always working, and you begin to lose the clarity and signal integrity. An experienced engineer tries to eliminate unnecessary amplifying circuits in the signal path—that's our approach here. The VCA should only act when it's needed.

Compressors and limiters are generally used while recording tracks as opposed to during mixdown, since one of the main benefits in compressing the signal is that you can get a more consistently hot signal on tape.

Listen to the different versions of the exact same vocal performance in Audio Examples 6-11 to 6-15. I've adjusted the level so that the peak of each version is at the same level. The only difference is the amount of compression. Pay special attention to the understandability of each word, the apparent tape noise, and the overall feel of each track.

..............................Audio Example 6-11

No Compression

..............................Audio Example 6-12

3-dB Gain Reduction

Chapter 6 ....PROCESSORS

253

Limiting

This graph represents a signal with a huge peak energy. Use a limiter on this kind of signal so the majority of the sound is unaffected (by the limiter's VCA) but the trouble spot is nearly eliminated.

The limiter can keep nearly any peak from overdriving the tape or from blasting through the mix. The results of effective limiting are often dramatic. If you start with a mix that has been level-impaired by a few quick blasts of energy, then you essentially remove those blasts, the entire mix level can be increased substantially, resulting in a much more powerful sound.

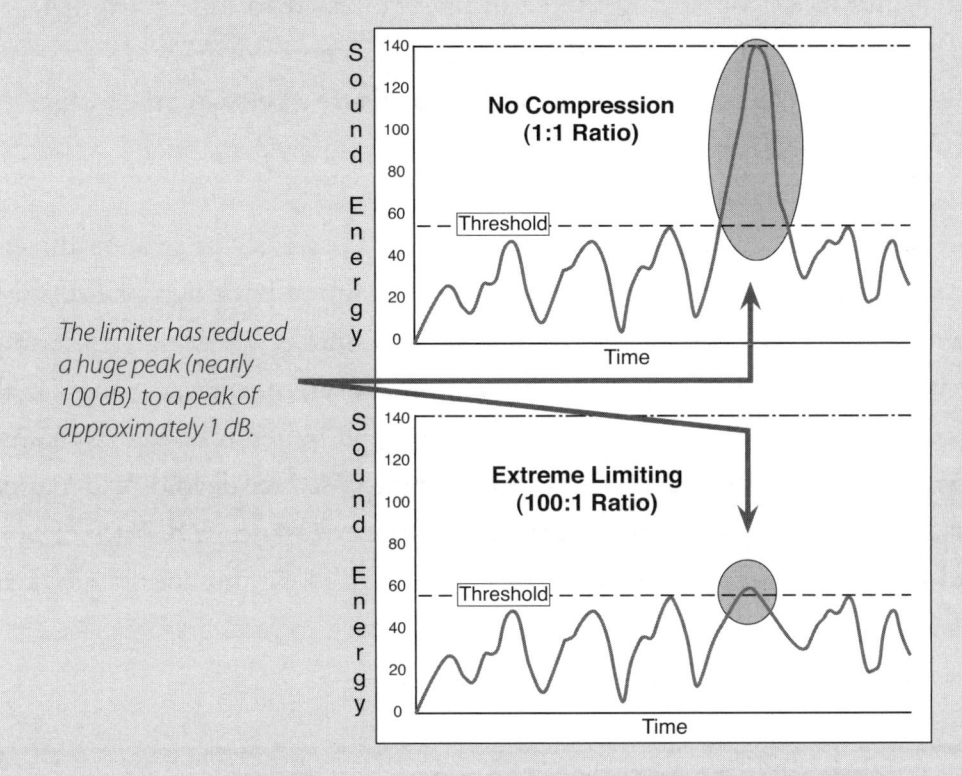

The limiter has reduced a huge peak (nearly 100 dB) to a peak of approximately 1 dB.

............................Audio Example 6-13

6-dB Gain Reduction

............................Audio Example 6-14

9-dB Gain Reduction

Audio Example 6-15 includes tape noise with the vocal. Listen to the compressor turning up and down (pumping and breathing).

254

Microphones & Mixers... by BILL GIBSON

····························· Audio Example 6-15

SHOULD I USE THE COMPRESSOR/LIMITER ON INPUT OR OUTPUT?

The compressor/limiter is typically used at the beginning of the signal path, just after the source enters the mixer input; however, there are valid reasons to incorporate this tool at each stage of the signal path. Compression and limiting provide an efficient means of controlling the level that's recorded to tape, hard disk, or any other analog or digital media. Recording instruments with a wide dynamic range often require constant level adjustments to ensure a consistently acceptable signal-to-noise ratio. These level changes can be performed manually, although they're typically much more reliable when performed electronically.

It's common to compress or re-compress audio tracks or groups during mixdown. Compressing the lead vocal or a stereo group of backing vocals during mixdown provides the engineer the opportunity to finely craft their positioning within the dynamic audio spectrum. However, avoid over-compressing any signal. Part of the power of audio rests in its dynamic content. When robbed of dynamic contrast, music and other audio sources lose impact. A mix devoid of dynamic contrast is tedious for the listener. Music is, at its very essence, a balance of tension and release. Sound that remains constant in amplitude and energy provides no release and eventually wears the listener out. Strive to find the best dynamic contrast for your music.

SETUP SUGGESTIONS FOR THE COMPRESSOR/LIMITER

- Adjust Ratio to determine function. Settings below 10:1 produce compression; settings from 10:1 to infinity:1 produce limiting.
- Set attack time fast or slow, depending on the audio source and desired effect.
- Set release time to about .5 seconds for general use.
- Select soft knee for gentle compression, or hard knee for limiting applications.
- Select RMS for most compression applications or peak for most limiting applications.

Chapter 6 ....PROCESSORS

255

+ Adjust the threshold for the amount of gain reduction that you want. You should typically have 3–6 dB of reduction at the strongest part of the track, and there should be times when there is no gain reduction.
+ Consider all rules carefully, then break them at will, and intentionally, anytime the music demands.

This is the textbook approach for the most natural and least audibly conspicuous compression.

If you've achieved 6 dB of gain reduction, you're able to boost your overall level to tape by 6 dB over what it would have been without the compressor. With the entire track boosted, we can hear the nuances and softer passages more clearly in the mix. As an additional bonus, the complete track (including the soft passages) will be 6 dB further away from the noise floor than it was before compression. Also, since the peak level has been increased by 6 dB, the vertical bit-depth resolution will benefit in the digital domain.

In most cases, compressors are essential tools for making professional-sounding audio recordings. If you are involved in audio for video and television, compressors are essential because of the limited dynamic range in these mediums.

HOW MUCH IS ENOUGH?

Though the musical and artistic needs of any recording must dictate the use and application of all available tools, there are some guidelines for most tasks which should be considered.

Compression and limiting are generally most effective when gain reduction occurs several times throughout a recording, yet most of the audio is left untouched—beneath the threshold. If the compressor/limiter is always turning the signal down and back up again, optimum gain reduction is not being achieved; in addition, adverse side effects called pumping and breathing occur.

PUMPING

Pumping is the result of the level-control circuitry reducing gain, as the amplitude exceeds the threshold, then turning back up again as the signal dips below the threshold. In some very "in-your-face" commercial pop music, a certain degree of pumping is acceptable to some artists; however, there are other ways to keep the

Microphones & Mixers.. by BILL GIBSON

mix optimally present without risking serious audio quality degradation. Pumping is often viewed as more extreme than breathing.

BREATHING

Breathing is like pumping, although the actual breathing sound is derived from a signal with a high noise content. As the noise, or room ambience, is decreased and increased as it crosses the threshold, it creates a sound similar to breathing. Whereas pumping and breathing are essentially the same technical anomaly, pumping is often associated with a full-range signal, and breathing is associated with an airy, high-frequency sound.

.........................Audio Example 6-16

More Pumping and Breathing

Listen to these example of pumping and breathing. Notice the difference in the effect as the audio content changes.

COMMERCIAL POP MUSIC

Commercial pop music is often highly compressed and limited. The fact that most highly commercial music is heard in listening environments with intrusive ambience motivates artists and producers to contain their recording an a very narrow dynamic range—the music must be heard over road noise, crowd noise, clanking glasses, breaking plates, and the "blue-light special" announcement. Most commercial recordings head immediately up to 0 VU and stay there through the duration of the song; this doesn't make for a very emotionally dynamic work of art, but it does produce music that consistently holds the listener's attention.

CLASSICAL RECORDING

Classical recordists typically use compression and limiting sparingly. The dynamic realism of a symphony or soloist—held in high regard throughout the classical listening community—is very integral to artistic expression. Too much dynamic processing changes the balance between ambience and music, altering the individualism of a highly-skilled ensemble or artist.

When included, dynamic processing is typically very subtle and understated. Limiting is sometimes used for extreme peaks, or gentle compression might be included to help smooth out the loudest sections; however, dynamic processing is almost always subtle and sonically inconspicuous.

THE PURIST AND COMPRESSION

The audio purist always strives to eliminate amplifying circuitry from the signal path. Therefore, many engineers prefer to control levels manually, using a mixer fader, to facilitate the most accurate, pristine, and natural-sounding recording. However, compressors and limiters have been so commonly used in commercial music recording that the sound achieved by their use has become expected; the impact, dynamic control, and punch that they provide has become, in some producer/engineers' opinions, sonically essential.

MULTIBAND COMPRESSOR/LIMITERS

Dynamic processing of a full-bandwidth signal presents a unique set of considerations. A single instrument or voice is typically functional in a specific frequency range; extreme bandwidth isn't a real consideration. In these instances, a full-bandwidth compressor/limiter is effective and preferred. On the other hand, a full-bandwidth recording (like a complete mix of a commercial pop song) contains an impressive amount of all frequencies. Low-frequency content contains the greatest amplitude, so, when running this type of signal through a full-range compressor/limiter, the low frequencies tend to activate the gain reduction circuitry first and most often. As the mix level changes in response to the bass frequencies, the high frequencies also change; they decrease during gain reduction, then they increase as the signal is turned back up.

. . . . Audio Example 6-17
Full-Range Recording through a Normal Compressor

Listen to this example of a full-range recording passing through a normal compressor/limiter. Notice how the high frequencies ride along with the lows, as gain reduction comes and goes.

A multiband compressor/limiter divides the audio into multiple frequency ranges (frequency bands). Each frequency band is compressed separately. Whereas the audible spectrum is typically divided into lows, mids, and highs, the actual crossover points, between bands, are often user-adjustable. Since each frequency range is dynamically controlled independent of the others, there is less audible pumping and breathing, and the mix stays more consistently in the forefront of the dynamic spectrum.

Most commercial mixes pass through some form of multiband compression eventually. Mastering engineers use these tools to help raise the overall level and impact of recordings so they'll sound loud and full in comparison with other professional recordings. Overuse of the multiband compressor can result in a lifeless sound; on the other hand, tastefully aggressive multiband dynamic control can result in a very punchy and impressive recording.

Multiband Compressor/Limiters

Multiband compression divides the audible spectrum into multiple bands, compressing each band separately. This typically provides a very punchy and powerful sound, although it offers the most potential for coloration as each band responds uniquely.

The Waves L3 multiband limiter, below, provides separate limiting for five user-selected bands. In addition, tools like these often provide level adjustment for each band, giving the user a chance to influence overall timbre.

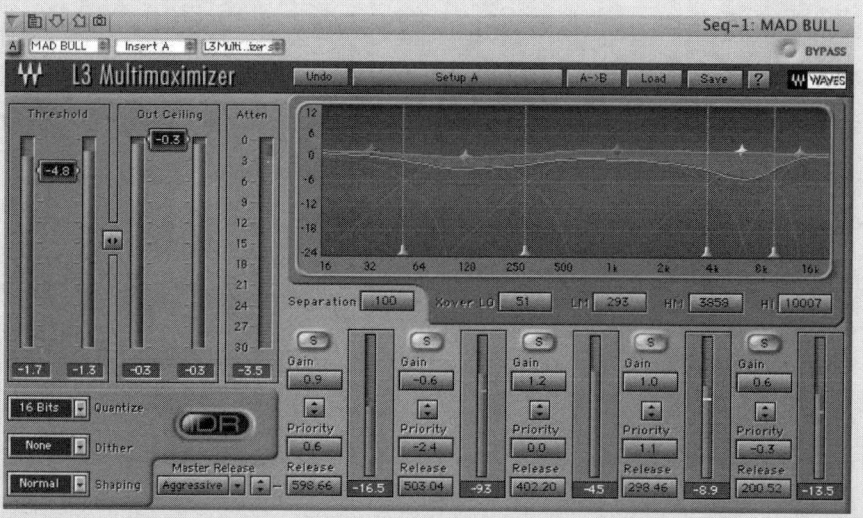

Chapter 6 ....PROCESSORS

259

Listen to this example of a full-bandwidth mix passing through a multiband compressor/limiter. Notice the change in sound as the processor switches in and out.

Be aware that, as the bands compress separately, the mix changes slightly. When the low-frequency gain is reduced, the highs and mids will probably stick out more in the mix; or, when the mids are reduced, the vocals might be suddenly buried. Careful selection of ratios and crossover points between bands will usually solve these problems, though these details do require your utmost attention.

.... Audio Example 6-18
Full-Range Recording through a Multiband Compressor

DE-ESSERS

A de-esser is a frequency-specific compressor, which reacts quickly to signals with strong high-frequency content—in particular, the common frequency range of the letters "s," "t," and "k." The purpose of the de-esser is to compensate for poorly compressed vocals. When a vocal track is recorded with the threshold too low and the attack time too slow, the initial transient sounds are overexaggerated, resulting in over-modulation of analog tape, over-stimulation of reverberation, and just a generally obnoxiously sibilant sound. Since the de-esser reacts quickly to high frequencies, it can usually solve a sibilance problem.

Most compressor/limiters can function as a de-esser. Simply select a fast attack time, patch the side chain to an external equalizer which has the frequencies between about 3 and 6 kHz boosted (depending on the sound being de-essed), select the side chain as the processor trigger, then adjust the threshold so gain reduction occurs whenever a transient problem occurs.

TUBE VERSUS SOLID-STATE COMPRESSOR/LIMITERS

Compressor/limiters over the years have been manufactured using vacuum tube circuitry, solid-state circuitry, and even a combination of both. Some engineers just love old equipment: They love the look, they love the history, they love the feel, and they love the sound. Realistically, high-quality compressors and limiters

Microphones & Mixers... by BILL GIBSON

are very useful, whether they use tube or solid-state technology. Listen to the tool, then choose the sound that provides the best support for the musical vision.

Tube technology typically sounds warmer and fuller than solid-state technology, especially as it's pushed to the limit of its capability. When a vacuum-tube audio circuit reaches distortion, the waveform is smoother and more rounded than a comparable solid-state waveform. In comparison, distortion of a solid-state circuit causes the waveform to be clipped off in an extreme way, creating a harsh and brittle sound. For this reason, musical styles that contain a wide dynamic range or very aggressive instrumentation and orchestration are often recorded using tube technology.

Listen to the difference in the sound between these compressors. Keep in mind that these are tools; tubes might be the best for one application while solid-state is better for another.

Solid-state technology is, in theory, far quieter than tube technology. Utilizing high-quality solid-state compressor/limiters is very desirable in a context where the amplifying circuitry is not being over taxed. They can be the most accurate and sonically appealing compressors and limiters, as long as the processor has plenty of headroom to avoid any kind of waveform distortion.

. . . . Audio Example 6-19
Sonic Comparison of Tube and Solid-State Compression

THE GATE/EXPANDER

The gate and expander are in the same family as the compressor/limiter. They're also centered on a VCA, and the VCA still turns the signal down. When the VCA is all the way up, the signal is at the same level as if the VCA weren't in the circuit—unity gain.

When the compressor/limiter senses the signal passing the threshold in an upward way, it turns down the signal that's above the threshold. The amount of gain reduction is determined by the ratio control. In contrast, when the gate/expander senses the signal passing the threshold in a downward way, the VCA turns the signal down even further. In other words, everything that's below the threshold is turned down.

Chapter 6 ....PROCESSORS

261

The Controls on a Gate/Expander

The controls on the gate/expander are essentially the same as the controls on a compressor/limiter. The threshold is the control that determines how much of the signal is acted on by the unit. The attack and release times do the same thing here that they did on the compressor: They control how quickly the unit acts once the signal has passed the threshold and how fast the unit turns the signal back up once the signal is no longer below the threshold.

The range control on the expander/gate correlates to the ratio control on the compressor/limiter. In fact, some multifunction dynamic processors use the same knob to control both ratio and range. The ratio on a compressor determines how far the VCA turns the signal down once it passes the threshold in an upward direction. The range on a gate/expander determines how far down the VCA will turn the signal once it passes the threshold in a downward direction.

When the signal gets below the threshold and the range setting tells the VCA to turn all the way off, the unit is called a gate. When the signal is below the

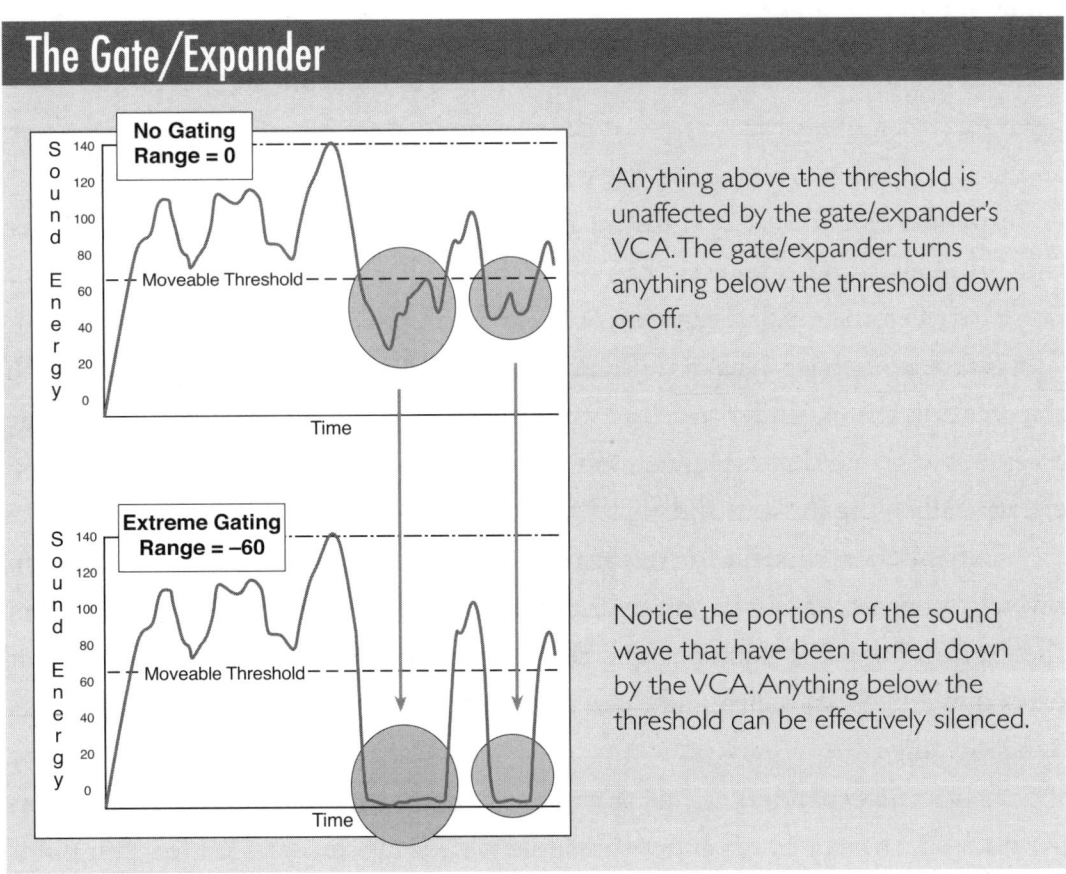

The Gate/Expander

Anything above the threshold is unaffected by the gate/expander's VCA. The gate/expander turns anything below the threshold down or off.

Notice the portions of the sound wave that have been turned down by the VCA. Anything below the threshold can be effectively silenced.

Microphones & Mixers... by BILL GIBSON

threshold, the gate is closed. The gate closes behind the sound and doesn't open again until the signal is above the threshold.

The range can also be adjusted so that the VCA only turns down the signal part of the way once it gets below the threshold. In this case, the unit is called an expander.

A gate is called a gate because it opens and closes when it senses the signal coming and going across the threshold. An expander is called an expander because it expands the dynamic range of the music. It creates a bigger difference between the softer and the louder sounds by turning the softer parts down.

The most common type of expander doesn't turn the louder parts up; it just seems to in relation to the softer parts. Specifically, an expander that turns the softer sounds down is called a downward expander. There is also an upward expander, which boosts the louder parts above unity. Upward expanders aren't very common and are somewhat noisy and difficult to control in a medium such as magnetic tape. Unless I specify otherwise, I'll refer to a downward expander throughout this course simply as an expander.

Gates Versus Expanders

Gates are especially useful in getting rid of noise, either from an instrument like a noisy electric guitar or from tape. If the threshold is set just above the noise floor, as the signal fades to the noise, the gate will simply continue the fade to silence. An expander can do exactly the same thing, but it will turn the noise down rather than off. Gates and expanders can really clean up a recording by getting rid of noise between the musical segments of each track.

Some units have a separate button to select the gate or the expander. Both the gate and the expander have the same controls, including full use of the range control. I've found that expanders are usually smoother in their level changes and are typically more musical and glitch free than gates.

Expanders are useful for restoring dynamic range to a signal that has been severely compressed. If the compressor reduced the loud parts by 9 dB, then, in theory, if the signal is expanded and the range control is adjusted to turn the soft parts down by 9 dB, we should have a pretty reasonable facsimile of our original dynamic range.

Gates and expanders are usually used on mixdown, rather than when recording tracks. If the gate or expander threshold were set incorrectly, some softer notes might not get printed to tape because they couldn't open the gate. If this happens, these softer notes are gone forever. Thus, the safest approach is to use gates and expanders on mixdown. During mixdown, the threshold can be non-destructively adjusted.

In a small setup, we often need to gate as we're recording (if we're going to gate at all) because of a lack of tracks and gates. This can work just fine, but more care must be taken in setting the processor, and the musical performance must be more consistent and predictable. For instance, a noisy guitar track is often gated during recording because the noise is consistent and it's easy to set the threshold so that the guitar sound comes through fine and the noise never touches tape.

So, a gate and an expander are really the same tool. The gate is an extreme version of an expander, with the gate turning the soft parts off where the expander just turns them down.

. . . . Audio Example 6-20
Guitar - No Gate

. . . . Audio Example 6-21
Guitar - Gated

. . . . Audio Example 6-22
Hi-Hat - No Expander

. . . . Audio Example 6-23
Hi-Hat - Expander

These dynamic range processors are all very useful and often essential in creating professional sounds. Each unit offers many creative and musical possibilities. As we study the individual instruments and their unique sound schemes, we'll use these processors time and again.

264

Microphones & Mixers.. by BILL GIBSON

CHAPTER TEST

1. Processors are typically connected to your system through:
    a.   channel inserts
    b.   aux sends
    c.   direct patch from an instrument
    d.   All of the above

2. If a performing group plays its loudest note at 110 dB and its softest note at 20 dB, its dynamic range is _____ dB.
    a.   110
    b.   90
    c.   130
    d.   45

3. The _____ is capable of responding to increases or decreases in input voltage by increasing or decreasing the output of the _____.
    a.   VCA, amplifying circuit
    b.   equalizer, highs and lows
    c.   compressor, limiter
    d.   delay, regeneration

4. Dynamic range processors are typically patched into the signal path of the microphone, instrument, or recorder track _____.
    a.   at the end of the signal path
    b.   in-line between the instrument and the console
    c.   via the aux bus
    d.   via the channel insert

5. When the compressor's VCA senses the signal exceeding a certain level, it acts on that signal and _____.
    a.   turns it back up
    b.   turns it down
    c.   reverses its phase
    d.   copies it

6. It's common to use minimal compression during tracking and then to use more extreme compression during mixdown.
    a.   True
    b.   False

7. The actual effect of compression primarily serves to:
    a.   turn the loud passages down
    b.   turn the soft passages and nuance up
    c.   squeeze the sound
    d.   shorten the sibilance

8. The attack time controls the amount of time it takes the compressor _____, once it's passed the threshold.
    a.   to finish attacking the level
    b.   to let go of the signal
    c.   to turn the signal down
    d.   to turn the nuance up

9. When compressing, fast release times work well with _____ to control peak levels. Slow release times work well with _____ to control average levels.
    a.   a compressor, a limiter
    b.   fast attack times, slow attack times
    c.   slow attack times, fast attack times
    d.   the threshold, the ratio

10. In compression, if the threshold is adjusted so that the loudest note of the song exceeds the threshold by 15 dB, and the ratio is 3:1, the 15-dB peak is reduced to a _____ dB peak—the gain is reduced by _____ dB.
    a.   12, 3
    b.   5, 10
    c.   45, 12
    d.   3, 1

Chapter 6 ....PROCESSORS

11. Ratio settings below 10:1 result in _____. Ratio settings of 10:1 and above result in _____.
    a. better intelligibility, a more musical mix
    b. limiting, compression
    c. a larger dynamic range, more clarity
    d. compression, limiting

12. If the VCA reduces the gain by 6 dB, the output/makeup gain control should be used to boost the output of the compressor/limiter by 6 dB.
    a. True
    b. False

13. Most of the time, there should be gain reduction. If there's not always gain reduction, the VCA isn't working, and you begin to lose the clarity and signal integrity.
    a. True
    b. False

14. _____ refers to average signal amplitude, based on the mathematical function of the _____.
    a. VU, +4 dBm
    b. Fletcher-Munson, Fletcher-Munson Curve
    c. RMS, Root Mean Square
    d. 0 VU, transient peak

15. The _____ is used to access control of the dynamics processor's VCA from an external source.
    a. key input
    b. side chain
    c. external input
    d. Both a and b
    e. All of the above

16. To meter gain reduction, some compressor/limiters use _____ and others use _____.
    a. peak levels, average levels
    b. a VCA, a DCO
    c. dynamics, triggers
    d. a series of LEDs, a VU meter

17. A limiter and compressor perform the same basic task, although a compressor typically controls level and amplitude in an extreme manner, while a limiter controls level and amplitude in a soft and gentle way.
    a. True
    b. False

18. _____ is the result of the level-control circuitry reducing gain as the amplitude exceeds the threshold, then turning back up again as the signal dips below the threshold.
    a. Breathing
    b. Pumping
    c. Pulsing
    d. Both a and b

19. When running a full-bandwidth signal through a full-range compressor/limiter, the _____ tend to activate the gain reduction circuitry first and most often.
    a. low frequencies
    b. high frequencies
    c. mid frequencies
    d. full-bandwidth signals

20. The _____ on a gate/expander determines how far down the VCA will turn the signal once it passes the threshold in _____ direction.
    a. ratio, a downward
    b. ratio, an upward
    c. threshold, the right
    d. range, a downward
    e. Both c and d

Test answers are on page 293

Effects

Effects processors add the third dimension to a mix. Room size and complexity are indicated by the way sound reacts in an acoustical space. The echoes and delays that happen after the original sound emanates from the source tell the brain what the surrounding environment is like. All of the effects processors (echoes, reverberation, and chorus effects) revolve around one thing: the delay.

WET VERSUS DRY

Wet and dry are two terms that refer to the amount of effected signal that is blended with the original dry signal. The relationship between wet and dry is quantified in a percentage; 100% wet refers to a signal that contains none of the original (dry) signal. A sound that is completely dry has none of the effect return combine with it (0% wet). An equal combination of the wet and dry signals is referred to as 50% wet.

PATCHING EFFECTS DEVICES

It's best to connect the output of your mixer's aux bus or effects send bus to the input of the effects unit. Next, connect the output of the effect to the mixer's effects return or into an available mixer channel.

Patching Effects Processors

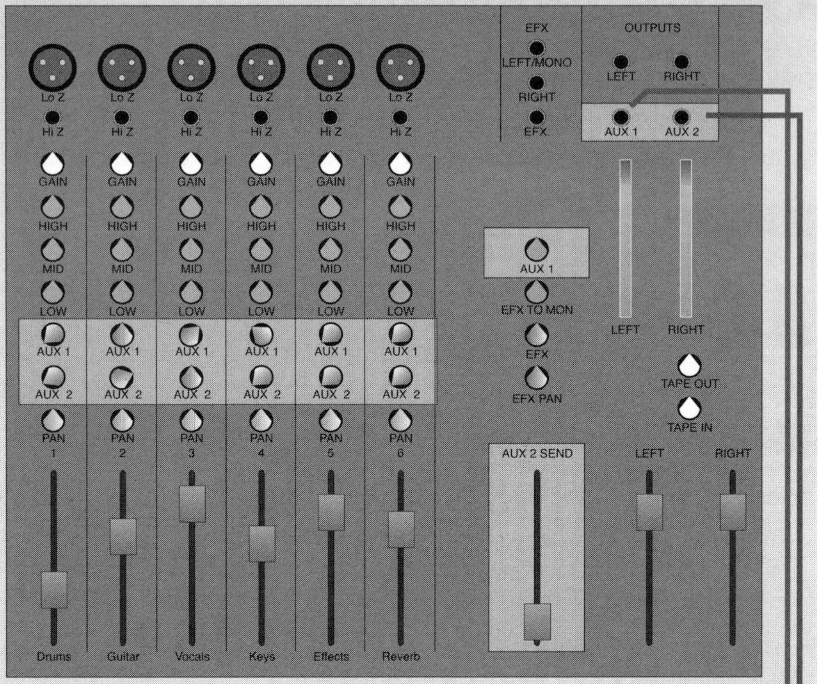

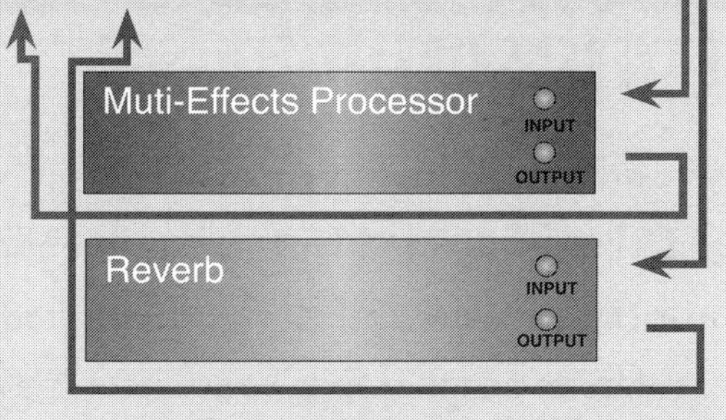

It's best to connect the output of your mixer's aux bus or effects send bus to the input of the effects unit. Next, connect the output of the effect to the mixer's effects return or into an available mixer channel.

When using effects, keep the original track dry, blending the 100% wet return with it for the best musical impact.

Muti-Effects Processor

INPUT

OUTPUT

Reverb

INPUT

OUTPUT

Most effects processors have a meter on the input for proper level adjustment, and many effects processors have a final output level adjustment.

When using effects it's always desirable to keep the original track dry and blend the 100% wet return with it for the best musical impact. In a small setup you might have to run the effects in-line, doing all of the blending from dry to wet within the effects unit. This can work well, but it's best to keep the dry and wet controls separate.

Chapter 7 .... EFFECTS

269

DELAY EFFECTS

A delay does just what its name says: It hears a sound and then waits for a while before it reproduces it. Current delays are simply digital recorders that digitally record the incoming signal, and then play it back with a time delay selected by the user. Delay parameters vary from unit to unit, but most delays have a range of delay length from a portion of a millisecond up to one or more seconds. This is called the delay time or delay length and is typically variable in increments of a millisecond.

Almost all digital delays are much more than simple echo units. Within the delay are all of the controls you need to produce slapback, repeating echo, doubling, chorusing, flanging, phase shifting, some primitive reverb sounds, and any hybrid variation you can dream up.

Slapback Delay

The simplest form of delay is called a slapback. The slapback delay is a single repeat of the signal. Its delay time is anything above about 35 ms. Any single repeat with a delay time of less than 35 ms is called a double.

To achieve a slapback from a delay, simply adjust the delay time and turn the delayed signal up, either on the return channel or on the mix control within the delay.

For a single slapback delay, feedback and modulation are set to their off positions. Slapback delays of between 150 ms and about 300 ms are very effective and common for creating a big vocal or guitar sound.

Audio Example 7-1 demonstrates a track with a 250 ms slapback delay.

· Audio Example 7-1

250 ms Slapback

Slapback delays between 35 and 75 ms are very effective for thickening a vocal or instrumental sound.

Audio Example 7-2 demonstrates a track with a 50 ms delay.

· Audio Example 7-2

50 ms Slapback

Microphones & Mixers .... by BILL GIBSON

Slapback delay can be turned into a repeating delay. This smooths out the sound of a track even more and is accomplished through the use of the regeneration control. This is also called feedback or repeat.

This control takes the delayed signal and feeds it back into the input of the delay unit, so we hear the original, the delay, and then a delay of the delayed signal. The higher you turn the feedback up, the more times the delay is repeated. Practi-

Calculating Delay Times

Delays are an important part of creating a professional-sounding mix. It's usually best if the delays are in time with the music—it helps reinforce the groove.

Calculating the delay time per beat is simple, especially if you're recording to a digital audio workstation, using the built-in sequencer as a click. Use this formula to find the length of one beat: 60,000 ÷ bpm. Here's the logic. If you keep it tucked away in your memory banks, you'll never need to look at a sheet of numbers in a grid again.

- There are 1,000 ms/second.

- There are 60 seconds/minute.

- Therefore, there are 60,000 ms/minute.

- Tempos are stated in beats per minute (bpm).

- Therefore, the total number of ms/minute (60,000) divided by the number of beats in a minute derives the number of ms per beat.

There are typically four beats per minute. Delays that work well, in support of the musical groove, are in time with the quarter note, eighth note, sixteenth note, or eighth- and sixteenth-note triplets.

To calculate these subdivisions of the beat, divide the ms/beat:

- By 1.5 to calculate the quarter note triplet value.

- By 2 to calculate the eighth note value.

- By 3 to calculate the eighth note triplet value.

- By 4 to calculate the sixteenth note value.

- By 6 to calculate the sixteenth note triplet value.

Chapter 7 .... EFFECTS

271

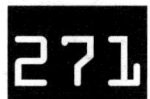

cally speaking, anything past about three repeats gets too muddy and does more musical harm than good.

The vocal track in Audio Example 7-3 starts with a simple single slapback, then the feedback raises until we hear three or four repeats.

····························· Audio Example 7-3
Repeating Delays

Why does a simple delay make a track sound so much bigger and better? Delay gives the brain the perception of listening in a larger, more interesting environment. As the delays combine with the original sound, the harmonics of each part combine in interesting ways. Any pitch discrepancies are averaged out as the delay combines with the original signal. If a note was sharp or flat, it's hidden when heard along with the delay of a previous note that was in tune. This helps most vocal sounds tremendously and adds to the richness and fullness of the mix.

The human brain gets its cue for room size from the initial reflections, or repeats, that it hears off surrounding surfaces. Longer delay times indicate to the brain that the room is larger. The slapback is really perceived as the reflection off the back wall of the room or auditorium as the sound bounces back (slaps back) to the performer. Many great lead vocal tracks have used a simple slapback delay as the primary or only effect. Frequently, this delay sounds cleaner than reverb and has less of a tendency to intrusively accumulate.

Slapback delay, often called echo, is typically related in some way to the beat and tempo of the song. The delay is often in time with the eighth note or sixteenth note, but it's also common to hear a slapback in time with the quarter note or some triplet subdivision. The delay time affects the rhythmic feel of the song. A delay that's in time with the eighth note can really smooth out the groove of the song, or if the delay time is shortened or lengthened just slightly, the groove may feel more aggressive or relaxed. Experiment with slight changes in delay time.

It's easy to find the delay, in milliseconds, for the quarter note in your song, especially when you're working from a sequence and the tempo is already available on screen. Simply divide 60,000 by the tempo of your song (in beats per minute). 60,000 ÷ bpm = delay time per quarter note in milliseconds (in Common time).

The slapback effect is often smoothed out by regenerating the delay, essentially creating multiple echoes or repeats. It's common to use a delay with two to five, or more, delays. This has a blending effect on most mixes.

Doubling/Tripling

Combining a single delay of less than 35 ms with the original track is called doubling. Combining two separate delays of less than 35 ms with the original track is called tripling. The short delay(s) can combine with the original track to sound like two people (or instruments) on the same part. Often, performers will actually record the same part two or three times to achieve the doubled or tripled sound, but sometimes the electronic simulation is quicker, easier, and sounds more precise. Audio Example 7-4 demonstrates an 11 ms delay (with no feedback and no modulation) combined with the original vocal. At the end of the example, the original and the delayed double pan apart in the stereo spectrum. This can be a great sound in stereo, but is a potential problem when summing to mono.

· Audio Example 7-4

11 ms Vocal Delay

When doubling, use prime numbers for delay times. You'll hear better results when your song is played in mono. A prime number can only be divided by one and itself (e.g., 1, 3, 5, 7, 11, 13, 17, 19, 23, 29, and so on).

Modulation

The modulation control on a delay is for creating chorusing, flanging, and phase shifting effects. The key factor here is the LFO (low-frequency oscillator); its function is to continually vary the delay time. The LFO is usually capable of varying the delay from the setting indicated by the delay time to half of that value and back. Sometimes the LFO control is labeled modulation.

As the LFO is slowing down and speeding up the delay, it's speeding up and slowing down the playback of the delayed signal. In other words, modulation actually lowers and raises the pitch in exactly the same way that a tape recorder does if the speed is lowered and raised. Audio Example 7-5 demonstrates the sound of the LFO varying the delay time. This example starts subtly, with the variation from the original going down slightly, then back up. Finally, the LFO varies dramatically downward, then back up again.

· Audio Example 7-5

The LFO

On most usable effects, these changes in pitch are slight and still within the boundaries of acceptable intonation, so they aren't making the instrument sound out of tune. In fact, the slight pitch change can have the effect of smoothing out any pitch problems on a track.

As the pitch is raised and lowered, the sound waves are shortened and lengthened. We know that when two waveforms follow the same path, they sum together. The result is twice the amount of energy. We also know that when two waveforms are out of phase, they work against and cancel each other, either totally or partially.

When the modulation is lengthening and shortening the waveform and the resulting sound is combined with the original signal, the two waveforms continually react together in a changing phase relationship. They sum and cancel at varying frequencies. The interaction between the original sound and the modulated delay can simulate the sound we hear when several different instrumentalists or vocalists perform together. Even though members of a choir try their hardest to stay in tune and together rhythmically, they're continually varying pitch and timing. These variations are like the interaction of the modulated delay with the original track. The chorus setting on an effects processor is simulating the sound of a real choir by combining the original signal with the modulated signal.

The speed control adjusts how fast the pitch raises and lowers. These changes might happen very slowly, taking a few seconds to complete one cycle of raising and lowering the pitch, or they might happen quickly, raising and lowering the pitch several times per second.

Audio Example 7-6 demonstrates the extreme settings of speed and depth. It's obvious when the speed and depth controls are changed here. Sounds like these aren't normally used, but when we're using a chorus, flanger, or phase shifter, this is exactly what is happening, in moderation.

. . . . Audio Example 7-6
Extreme Speed and Depth

Phase Shifter

Now that we're seeing what all these controls do, it's time to use them all together. Obviously, the delay time is the key player in determining the way that the depth and speed react. If the delay time is very, very short, in the neighborhood of 1

274

Microphones & Mixers.. by BILL GIBSON

ms or so, the depth control will produce no pitch change. When the original and affected sounds are combined, we hear a distinct sweep that sounds more like an EQ frequency sweeping the mids and highs. With these short delay times, we're really simulating waveforms, moving in and out of phase, unlike the larger changes of singers varying in pitch and timing. The phase shifter is the most subtle, sweeping effect, and it often produces a swooshing sound.

Audio Example 7-7 demonstrates the sound of a phase shifter.

· Audio Example 7-7

Phase Shifter

Flanger

A flanger has a sound similar to the phase shifter, except it has more variation and color. The primary delay setting on a flanger is typically about 20 ms. The LFO varies the delay from near 0 ms to 20 ms and back, continually. Adjust the speed to your own taste.

Flangers and phase shifters work very well on guitars and Rhodes-type keyboard sounds. They provide a rich blend and interesting harmonic motion.

Audio Example 7-8 demonstrates the sound of a flanger.

· Audio Example 7-8

Flanger

Chorus

The factor that differentiates a chorus from the other delay effects is, again, the delay time. The typical delay time for a chorus is about 15 to 35 ms, with the LFO and speed set for the richest effect for the particular instrument voice or song. With these longer delay times, as the LFO varies, we actually hear a slight pitch change. The longer delays also create more of a difference in attack time. This also enhances the chorus effect. Since the chorus gets its name from the fact that it's simulating the pitch and time variations that exist within a choir, it might seem obvious that a chorus works great on background vocals. It does. Chorus is also an excellent effect for guitar and keyboard sounds.

Audio Example 7-9 demonstrates the sound of a chorus.

Chapter 7 .... EFFECTS

275

.... Audio Example 7-9

Chorus

Phase Reversal and Regeneration

The regeneration control can give us multiple repeats by feeding the delay back into the input so that it can be delayed again. This control can also be used on the phase shifter, chorus, and flange. Regeneration, also called feedback, can make the effect more extreme or give the music a sci-fi feel. As you practice creating these effects with your equipment, experiment with feedback to find your own sounds.

Most units have a phase reversal switch that inverts the phase of the affected signal. Inverting the phase of the delay can cause very extreme effects when combined with the original signal (especially on phase shifter and flanger effects). This can make your music sound like it's turning inside out.

Audio Example 7-10 starts with the flanger in phase. Notice what happens to the sound as the phase of the effect is inverted.

.... Audio Example 7-10

Inverting Phase

Delay Settings for Various Delay Effects

Effect	Delay A	Delay B (Stereo)	LFO	Speed	Regener-ation	Phase
Slapback	35–350ms		No	No	No	No
Echo (Repeats)	35–350ms		No	No	2–10	No
Reverb	15–35ms	15–35ms	No	No	Several	No
Doubling	1–35ms		No	No	No	No
Tripling	1–35ms	1–35 ms	No	No	No	No
Phase Shifter	0.5–2ms	0.5–2 ms	Yes	Low	Medium	Yes/No
Flanger	10–20ms	10–20ms	Yes	Low	Medium	Yes/No

276

Microphones & Mixers.. by BILL GIBSON

Stereo Effects

The majority of effects processors are stereo, and with a stereo unit, different delay times can be assigned to the left and the right sides. If you are creating a stereo chorus, simply set one side to a delay time between 15 and 35 ms, then set the other side to a different delay time, between 15 and 35 ms. All of the rest of the controls are adjusted in the same way as a mono chorus. The returns from the processor can then be panned apart in the mix for a very wide and extreme effect. Listen as the chorus in Audio Example 7-11 pans from mono to stereo.

. . . . Audio Example 7-11

Stereo Chorus

For a stereo phase and flange, use the same procedure. Simply select different delay times for the left and right sides.

Understanding what is happening within a delay is important when you're trying to shape sounds for your music. Sometimes it's easiest to bake a cake by simply pressing the Bake Me a Cake button, but if you are really trying to create a meal that flows together perfectly, you might need to adjust the recipe for the cake. That's what we need to do when building a song, mix, or arrangement; we must be able to custom fit the ingredients.

REVERBERATION EFFECTS

As we move from the delay effects into the reverb effects, we must first realize that reverb is just a series of delays. In fact, modern reverberation devices are capable of all delay effects. However, some devices are limited to producing either delay or reverberation effects. Also, many software plug-ins specifically focus on a single function, often in emulation of classic hardware.

Reverberation is simulation of sound in an acoustical environment, like a concert hall, gymnasium, or bedroom. No two rooms sound exactly alike. Sound bounces back from all the surfaces in a room to the listener or the microphone. These bounces are called reflections. The combination of the direct and reflected sound in a room creates a distinct tonal character for each acoustical environment. Each one of the reflections in a room is like a single delay from a digital delay. When it bounces around the room, we get the effect of regeneration. When we

Chapter 7 .... EFFECTS

277

take a single short delay and regenerate it many times, we're creating the basics of reverberation.

Reverb must have many delays and regenerations working together in the proper balance, combining to create a smooth and appealing room sound.

Envision thousands of delays bouncing (reflecting) off thousands of surfaces in a room and then back to you, the listener—that's what's happening in the reverberation of a concert hall or any acoustical environment. There are so many reflections happening in such a complex order that we can no longer distinguish individual echoes.

Slapback Delay and Reflections

Sound travels at the rate of about 1,126 ft./sec. To calculate the amount of time (in seconds) it takes for sound to travel a specific distance, divide the distance (in feet) by 1,126 (ft./sec.): time = distance (ft.) ÷ speed (1,126 ft./sec.).

In a 100' long room, sound takes about 89 ms to get from one end to the other (100÷1,126). A microphone at one end of this room wouldn't pick up the slapback until it completed a round trip (about 178 ms after the original sound).

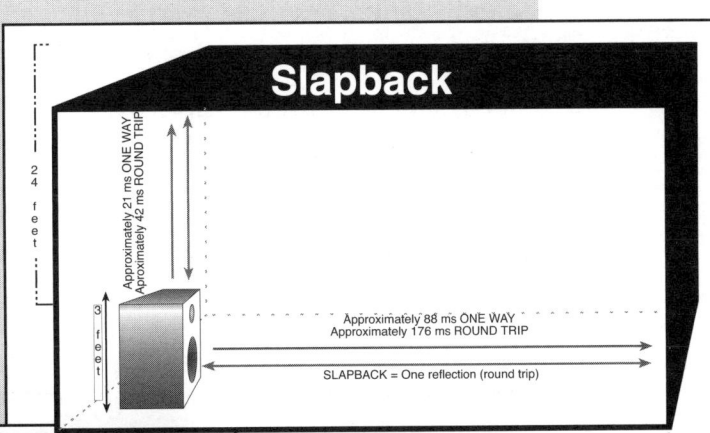

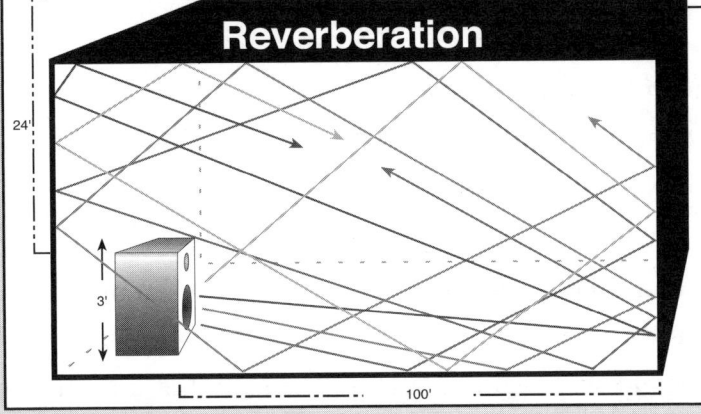

278

Microphones & Mixers.. by BILL GIBSON

Accurate and believable digital simulation is accomplished by producing enough delays and echoes to imitate the smooth sound of natural reverb in a room. The reason different reverb settings sound unique is because of the different combinations of delays and regenerations. The mathematical calculation and relations of the delays involved in a reverberation sound are called an algorithm.

A digital reverb is capable of imitating a lot of different acoustical environments and can do so with amazing clarity and accuracy. The many different echoes and repeats produce a rich and full sound. Digital reverbs can also shape many special effects that would never occur acoustically. In fact, these sounds can be so fun to listen to that it's hard not to overuse reverb.

Keep in mind that sound perception is not just two dimensional, left and right. Sound perception is at least three dimensional, with the third dimension being depth (distance). Depth is created by the use of delays and reverb. If a sound (or a mix) has too much reverb, it loses the feeling of closeness or intimacy and sounds like it's at the far end of a gymnasium. Use enough effect to achieve the desired results, but don't overuse effects.

Most digital reverberation devices offer several different sounds. These are usually labeled with descriptive names like halls, plates, chambers, rooms, etc.

Hall Reverb

Hall indicates a concert hall sound. These are the smoothest and richest of the reverb settings, with complex, long delay times that blend together to form a smooth decay over time. Typical hall algorithms have a decay time longer than 2 seconds, although user-adjustable controls allow for unnatural settings on hall sounds or any of the basic sounds.

Chamber Reverb

Chambers imitate the sound of an acoustical reverberation chamber, sometimes called an echo chamber. Acoustical chambers are fairly large rooms with hard surfaces. Music is played into the room through high-quality, large speakers, and then a microphone in the chamber is patched into a channel of the mixer as an effects return. Chambers aren't very common now that technology is giving us great sounds without taking up so much real estate. The sound of a chamber is smooth, like the hall's, but has a few more mids and highs.

Plate Reverb

Plates are inherently the brightest sounding of the reverbs. These sounds imitate a physical plate reverb. A true plate is a large sheet of metal (about 4' by 8') suspended in a box and allowed to vibrate freely. A speaker attached to the plate itself induces sound onto the plate. Two contact microphones are typically mounted on the plate at different locations to provide a stereo return. The sound of a true plate reverb has lots of highs and is very clean and transparent.

Room Reverb

A room setting imitates many different types of rooms that are typically smaller than hall or chamber sounds. These can range from a bedroom to a large conference room, or from a small bathroom with lots of towels to a large bathroom with lots of tile.

Rooms with lots of soft surfaces have little high-frequency content in their reverberation. Rooms with lots of hard surfaces have lots of high-frequency content in their reverberation.

Reverse Reverb

Most modern reverbs include reverse or inverse reverb. These are simply backwards-sounding reverbs. After the original sound is heard, the reverb swells and stops. It is turned around. These can actually be fairly effective if used in the appropriate context.

Gated Reverb

Gated reverbs have a sound that is very intense for a period of time, and then closes off quickly. They offer a very big sound without overwhelming the mix.

Though at one time this was a trendy, popular sound, the technique has been around for a long time. The original gated reverb sound actually used a room mic, distant from the source, in a large room patched through a gate. The trigger for the gate to open was set to the side chain, where a mic close to the source was patched. This was common on snare drum at one time. There was a close mic on the snare patched to the mixer and also patched into the trigger input of the gate, so when the snare was hit, the gate opened and you could hear the large sound of the room microphone(s). When the snare wasn't being hit, the room mics were off.

Microphones & Mixers.. by BILL GIBSON

Other Variations of Reverberation

There are many permutations of the reverberation sounds. You might see bright halls, rich plates, dark plates, large rooms, small rooms, or a bright phone booth, but they can all be traced back to the basic sounds of halls, chambers, plates, and rooms.

These sounds often have adjustable parameters. They let us shape the sounds to our music so that we can use the technology as completely as possible to enhance the artistic vision. We need to consider these variables so that we can customize and shape the effects.

· ·Audio Example 7-12

Reverberation Variations

· ·Video Example 7-1

Various Acoustic Space Recordings

EFFECTS PARAMETERS

Predelay

Predelay is a time delay that happens before the reverb is heard. This can be a substantial time delay (up to a second or two) or just a few milliseconds. The track is heard clean (dry) first, so the listener gets more of an up-front and close feel, then the reverb comes along shortly thereafter to fill in the holes and add richness.

Diffusion

Diffusion controls the space between the reflections. A low diffusion is equated with a very grainy photograph. We might even hear individual repeats in the reverb. A high diffusion is equated with a very fine-grain photograph, and the sound provides a very smooth wash of reverb.

Decay Time

Reverberation time, reverb time (RT), and decay time all refer to the same thing. Traditionally, reverberation time is defined as the time it takes for the sound to decrease to one-millionth of its original sound pressure level. In other words, it's the time it takes for the reverb to go away.

Chapter 7 .... EFFECTS

Decay time can typically be adjusted from about 1/10 of a second up to about 99 seconds. We have ample control over the reverberation time.

Density

The density control adjusts the initial short delay times. Low density is good for smooth sounds like strings or organ. High density works best on percussive sound.

.... Audio Example 7-13

Reverberation Parameters

Diffusion

Reverberation with low diffusion often sounds grainy because the reflections that make up the reverb sound are relatively far apart. There are some reverberation algorithms that sound good with lower diffusion, especially in the context of an up-tempo musical work. Low-diffusion reverb sounds don't usually work well in a very open ballad.

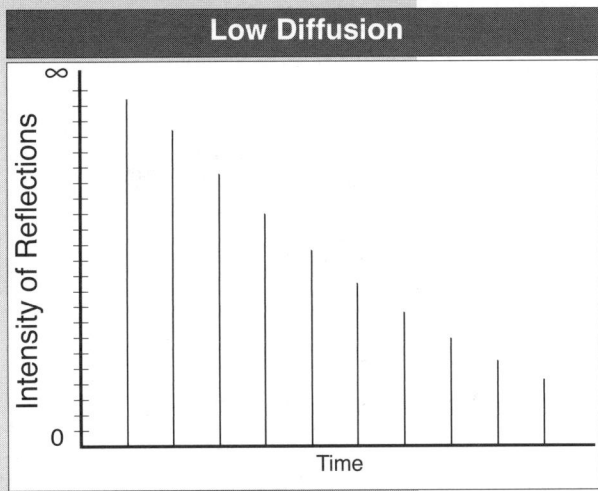

High-diffusion usually results in a smooth-sounding reverb. Since the reflections are close together in time, they blend together to create a wash of reverberation. If the arrangement or orchestration is very busy and the reverberation has high diffusion, the mix might sound too thick. High diffusion works well on ballads and very exposed arrangements.

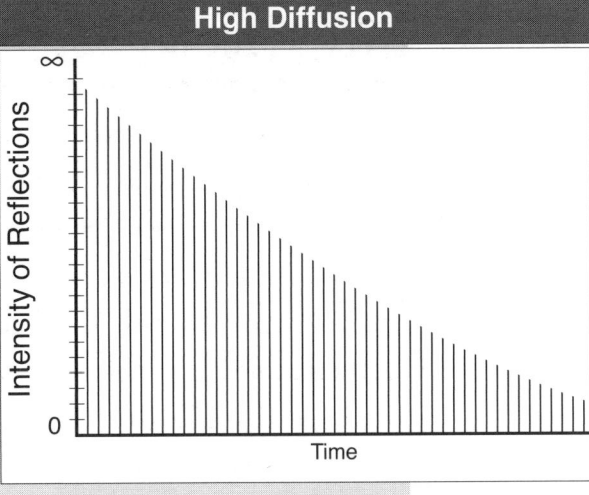

282

Microphones & Mixers.. by BILL GIBSON

Density

Low Density

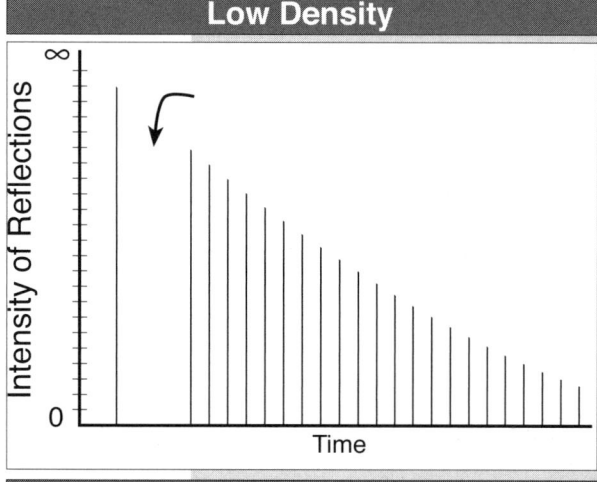

Low-density reverberation is good for smooth sounds like a string pad or a mellow organ. The increased spacing of the initial reflections can sound grainy on sounds with percussive attacks.

High Density

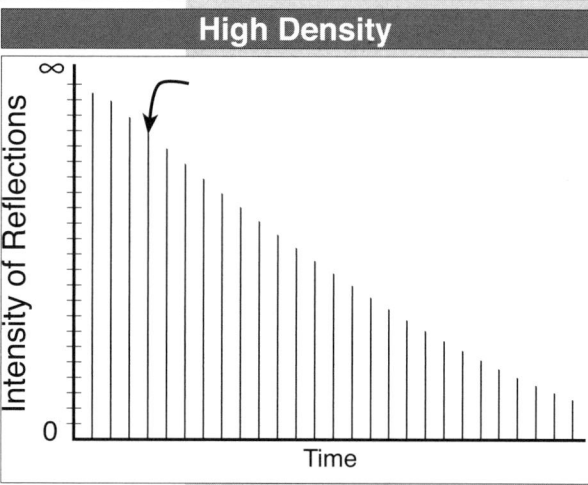

High-density reverberation sounds good on percussive instruments. The closer spacing of the initial reflections produces a smooth-sounding reverb on percussive attacks.

EFFECTS PLUG-INS

There are so many options when it comes to effects plug-ins! The key, at this point, is that you understand the effects parameters and the sonic impact they have. The parameters are the same across the board, whether you're recording with hardware or software. The nice thing about software plug-ins is that they typically have infinite control over every imaginable parameter. There is nothing between you and your creative options. That fact that there is an attractive visual interface for each software plug-in also aids in the understanding of how the effects are created, altered, and used.

Chapter 7 .... EFFECTS

SUMMARY

In the recording world of yesteryear the only way to adjust reverberation time was to dampen or release the springs in the spring reverb tanks or physically move a bar that moved a felt pad onto, or off of, the plate reverb. Trying to control reverb time in a true reverberation chamber is even more difficult. Current technology provides a myriad of variables when shaping reverb sounds. In fact, when you consider the number of possible options, it is mind boggling. We can design unnatural hybrids like a large room with a very short decay time and plenty of high frequency, or any other natural or unnatural effect.

Each parameter is important, and as we deal with individual guitar, drum, keyboard, and vocal sounds, reverb is a primary consideration.

Modern-day multi-effects processors typically sound great and offer countless control options. When you're selecting any device that provides reverberation effects, listen to it before you buy it. Some manufacturers really have the algorithms that sound great; other manufacturers wish they did.

Don't overlook the classic devices. They still sound good and provide unique creative options that are waiting to be rediscovered.

Microphones & Mixers... by BILL GIBSON

CHAPTER TEST

1. It is best to operate a reverberation device at _____ in order to get the best-sounding blend of _____ in the final mix.
   a. 50% wet, wet and dry audio
   b. 100% wet, wet and dry audio
   c. full digital level, effects
   d. 0% wet, the tracks

2. From your mixer, it's best to connect the output of _____ to the input of an effects device, and then to connect the output of the effects device to _____.
   a. an aux bus, the mixer's effects return
   b. an effects send, an available mixer channel
   c. an insert, an available mixer channel
   d. Both a and b
   e. All of the above

3. The simplest form of delay is called a _____. It is a _____ of the signal with a delay time that's greater than _____.
   a. slapback, single repeat, 35 ms
   b. reflection, replica, 100 ms
   c. simpleton, bounce back, the original
   d. predelay, ping pong, 35 ms

4. In order to calculate the delay setting that matches the 16th note, use the formula:
   a. bpm x 60 ÷ 4
   b. 60,000 ÷ bpm
   c. 60 x 1,000 ÷ bpm ÷ 2
   d. 60,000 ÷ bpm ÷ 4

5. Combining a single delay of less than 35 ms with the original track is called _____.
   a. doubling
   b. slapback
   c. tripling
   d. reflecting

6. The _____ on a delay is for creating chorusing, flanging, and phase shifting effects.
   a. regeneration control
   b. sweep generator
   c. modulation control
   d. LFO
   e. Both c and d

7. Considering the delay times used to create chorus, flanger, and phase shifter effects, which of the following answers lists them in ascending order?
   a. chorus, flanger, phase shifter
   b. chorus, phase shifter, flanger
   c. phase shifter, flanger, chorus
   d. flanger, phase shifter, chorus

8. When creating phase shifter effects, _____ of the delay can cause very extreme effects when combined with the original signal.
   a. increasing the regeneration
   b. inverting the phase
   c. varying the length
   d. All of the above

9. To create a stereo chorus, set one side to a delay time of _____, then set the other side to a delay time of _____.
   a. 17 ms, 23 ms
   b. 15 ms, 34 ms
   c. 11 ms, 49 ms
   d. Either a or b
   e. Any of the above

Chapter 7 .... EFFECTS

285

10. The combination of the _____ and _____ sound in a room creates a distinct tonal character for each acoustical environment.
    a.   direct, reflected
    b.   ambient, defracted
    c.   absorbed, diffused
    d.   direct, in-phase

11. In a 70'-foot long room, sound takes about _____ to get from one end to the other. A microphone at one end of this room wouldn't pick up the slapback until about _____ after the original sound.
    a.   70 ms, 105 ms
    b.   35 ms, 70 ms
    c.   62.5 ms, 125 ms
    d.   125 ms, 250 ms

12. A plate reverb has a sound that is very intense for a period of time, and then closes off quickly.
    a.   True
    b.   False

13. _____ are inherently the brightest sounding of the reverbs.
    a.   Halls
    b.   Chambers
    c.   Plates
    d.   Rooms

14. _____ is a time delay that happens before the reverb is heard.
    a.   Predelay
    b.   Delay
    c.   Predecay
    d.   RT

15. Depth is created by the use of _____ and _____.
    a.   panning, mixing
    b.   ambience, chorus effects
    c.   delays, reverberation
    d.   EQ, reverberation

16. Reverberation time is defined as the time it takes for the sound to decrease to _____ of its original sound pressure level.
    a.   1%
    b.   half
    c.   10%
    d.   one-millionth

17. When adjusting reverberation parameters, the _____ control adjusts the initial short delay times.
    a.   diffusion
    b.   density
    c.   delay length
    d.   decay time

18. When adjusting reverberation parameters, _____ is equated with a very grainy photograph.
    a.   low diffusion
    b.   low density
    c.   high density
    d.   high diffusion

19. The _____control turns a single delay into multiple delays.
    a.   multiplier
    b.   LFO
    c.   regeneration
    d.   All of the above

20. Effects plug-ins utilize the same parameters as traditional hardware devices. Once you understand the parameters involved in shaping and building effects, it shouldn't matter whether you're using hardware or software.
    a.   True
    b.   False

Test answers are on page 293

Index

Test Answers

Chapter 1

1. .... c.
2. .... e.
3. ....b
4. .... a
5. .... a
6. c
7. ....d
8. ....b
9. c
10. ....d
11. ....a
12. c
13. ....d
14. c
15. ....d
16. ....b
17. ....a
18. ....d
19. ....d
20. .... e

Chapter 2

1. c
2. ....d
3. ....b
4. ....d
5. .... e
6. ....b
7. .... a
8. ....b
9. ....b
10. ....d
11. ....a
12. ....b
13. ....b
14. ....b
15. ....d
16. ....b
17. c
18. ....b
19. ....a
20. ....b

Chapter 3

1. ....b
2. .... a
3. c
4. .... b
5. c
6. ....d
7. ....d
8 .... a
9. ....d
10. .... a
11. ....d
12. .... e
13. .... a
14. .... a
15. c
16. ....b
17. .... a
18. .... e
19. ....d
20. .... a

Chapter 4

1. ....b
2. c
3. .... e
4. .... a
5. .... a
6. .... a
7. .... b
8. ....b
9. ....b
10. .... a
11. ....d
12. c
13. c
14. ....b
15. .... a
16. ....b
17. ....d
18. ....b
19. ....d
20. ....d

Chapter 5

1. ....d
2. c
3. ....d
4. ....b
5. .... a
6. ....b
7. ....b
8. ....b
9. .... a
10. .... e
11. ....b
12. c
13. ....d
14. .... a
15. ....b
16. ....d
17. .... e
18. ....b
19. ....b
20. .... a

Chapter 6

1. ....d
2. ....b
3. .... a
4. ....d
5. ....b
6. .... a
7. ....b
8. c
9. ....b
10. ....b
11. ....d
12. .... a
13. ....b
14. c
15. .... e
16. ....d
17. ....b
18. ....d
19. .... a
20. ....d

Chapter 7

1. ....b
2. ....d
3. .... a
4. ....d
5. .... a
6. .... e
7. c
8. ....d
9. ....d
10. .... a
11. c
12. ....b
13. c
14. .... a
15. c
16. ....d
17. ....b
18. .... a
19. c
20. .... a

NOTES:

NOTES: _____

_____

_____

_____

_____

_____

_____

_____

_____

_____

_____

_____

_____

_____

_____

_____

_____

_____

_____

_____

_____

_____

_____

_____

The Entire Recording Process *Explained!*

The Hal Leonard Recording Method
by Bill Gibson is the first professional
multimedia recording method to take readers
from the beginning of the signal path to the final master mix.

ALL BOOKS INCLUDE  DVD

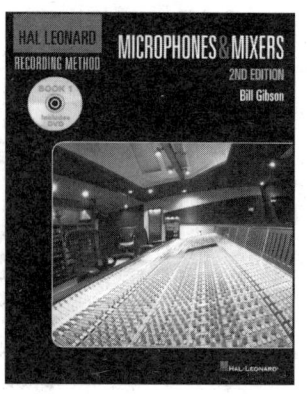

BOOK 1: MICROPHONES & MIXERS – 2ND EDITION
Revised and Updated

Topics include how professional microphones work, which to choose and why (plus accepted techniques for using them), understanding the signal path from mics to mixers and how to operate these critical tools to capture excellent recordings, and explanations of the most up-to-date tools and techniques involved in using dynamics and effects processors.

HL00333253 Book/DVD Pack $39.99
978-1-45840-296-7

BOOK 4: SEQUENCING SAMPLES & LOOPS

Learn to create amazing musi[c] productions using the lates[t] sequencing techniques with sample[s] and pre-recorded loops. Wit[h] detailed screen shots, illustration[s] video and audio examples, an[d] more on the accompanying DV[D] you're on your way to rounding o[ut] your recording education.

HL00331776 Book/DVD Pack ... $39.9[9]
978-1-4234-3051

BOOK 2: INSTRUMENT & VOCAL RECORDING – 2ND EDITION
Revised and Updated

This edition addresses new equipment and software concerns that affect the way excellent recordings are made. You'll learn what you need to know about capturing the best vocal and instrument tracks possible, no matter what kind of studio you are working in or what kind of equipment is used.

HL00333250 Book/DVD Pack $39.99
978-1-45840-292-9

BOOK 5: ENGINEERING & PRODUCING

Learn what it takes to engineer [a] session like a pro, combining t[he] teaching of the previous books [to] record excellent tracks ready [for] the mix. This book also dives in[to] what makes a producer great, h[ow] to inspire awesome performanc[es] from the musicians you reco[rd,] tricks for selecting the best trac[ks] from your sessions, and what [it] takes to be both producer a[nd] engineer on the same session.

HL00331777 Book/DVD Pack ... $39[.99]
978-1-4234-305[

BOOK 3: RECORDING SOFTWARE & PLUG-INS

Once you've learned how to use microphones and mixers, and to record instruments and vocals, you'll need to know how recording software programs work and how to choose and optimize your recording system. This book and DVD use detailed illustrations and screen shots, plus audio and video examples, to give you a comprehensive understanding of recording software and plug-ins.

HL00331775 Book/DVD Pack $39.95
978-1-4234-3050-6

BOOK 6: MIXING & MASTERING – 2ND EDITION
Revised and Updated

This important second edition de[m]onstrates techniques and pro[ce]dures that result in a polished [ ] and powerful master record[ing] using current plug-ins, softwa[re] and hardware. You'll then learn [how] to prepare the mastered rec[ord]ing for CD replication, stream[ing] or download. Updated illustrati[ons] photographs, and audio and vi[deo] examples on the accompany[ing] DVD will reinforce your underst[and]ing of what you need to mix [and] master like the pros.

HL00333254 Book/DVD Pack ... $3[
978-1-45840-2[

HAL•LEONARD®
www.halleonardbooks.com